MEMC
TECHNOLOGY IS BUILT ON US

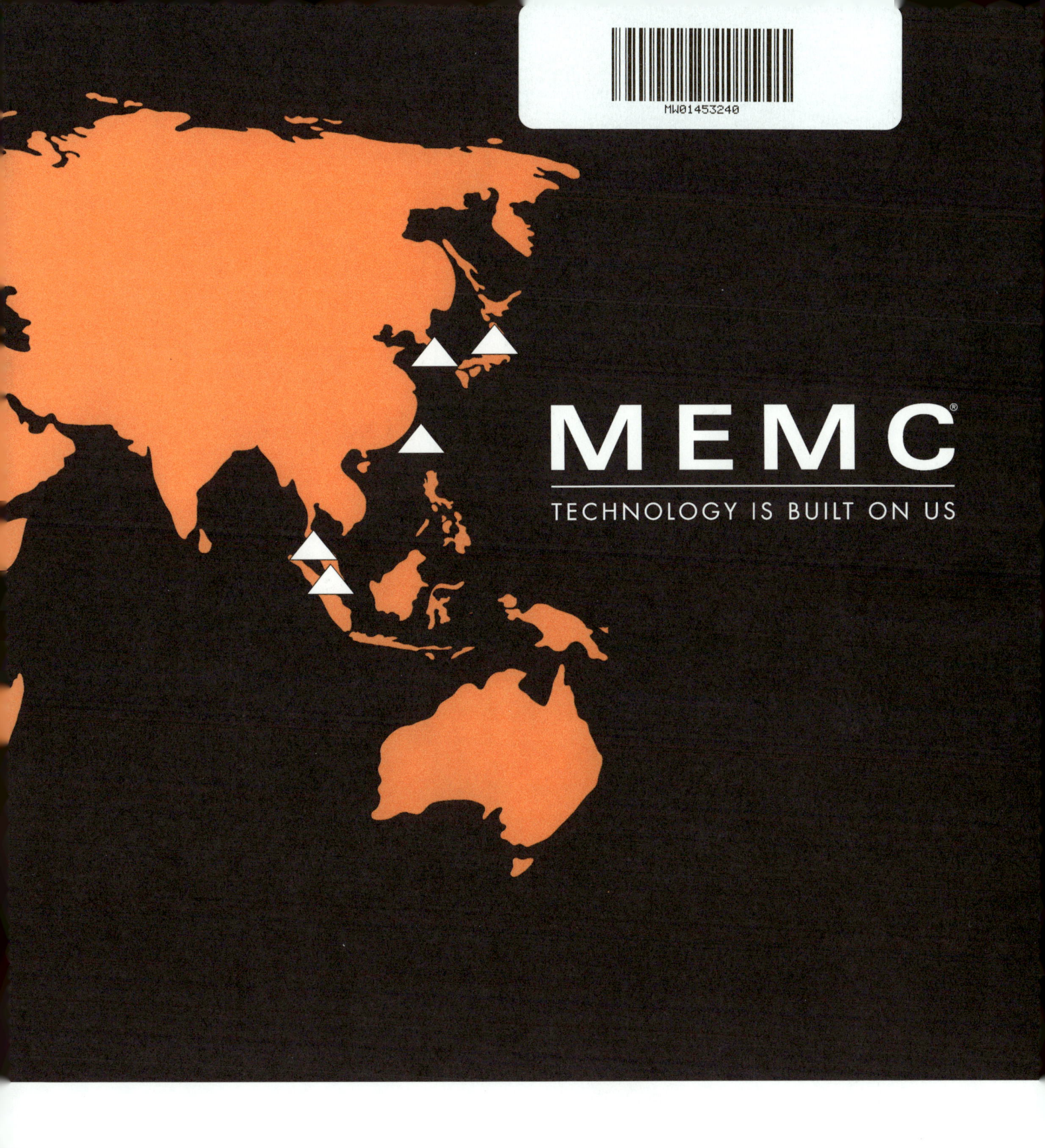

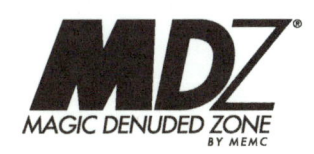

SILICON WORKSHOP **Optia** **aegis** **MDZ** MAGIC DENUDED ZONE BY MEMC

BUILDING ON THE PAST, READY FOR THE FUTURE:

A FIFTIETH ANNIVERSARY CELEBRATION OF MEMC ELECTRONIC MATERIALS, INC.

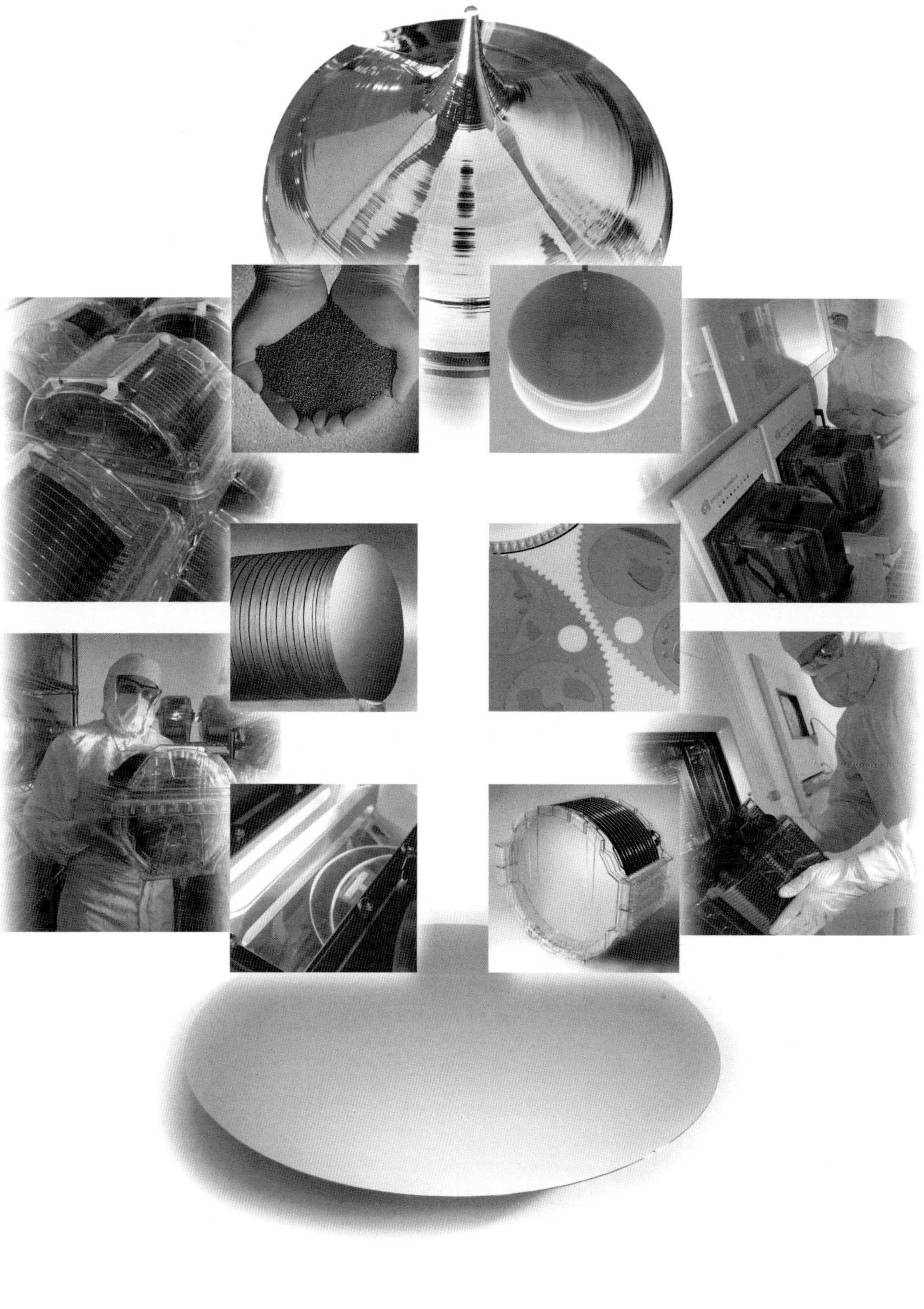

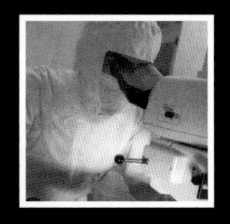

BUILDING ON THE PAST, READY FOR THE FUTURE:
A FIFTIETH ANNIVERSARY CELEBRATION OF MEMC ELECTRONIC MATERIALS, INC.

by Pat Swinger
Edited and contributions by Sheila O'Connell

THE
DONNING COMPANY
PUBLISHERS

Copyright © 2009 by MEMC

All rights reserved, including the right to reproduce this work in
any form whatsoever without permission in writing from the publisher,
except for brief passages in connection with a review.
For information, please write:

The Donning Company Publishers
184 Business Park Drive, Suite 206
Virginia Beach, VA 23462

Steve Mull, General Manager and Project Director
Barbara Buchanan, Office Manager
Pamela Koch, Senior Editor
Tonya Hannink, Graphic Designer and Marketing Specialist
Derek Eley, Imaging Artist
Pamela Engelhard, Marketing Advisor

Library of Congress Cataloging-in-Publication Data

Swinger, Patricia, 1951-
  Building on the past, ready for the future : a fiftieth anniversary celebration of MEMC Electronic Materials, Inc. / by Pat Swinger ; edited and contributions by Sheila O'Connell.
      p. cm.
  Includes bibliographical references.
  ISBN 978-1-57864-574-9 (hbk. : alk. paper)
 1. MEMC Electronic Materials—History. 2. Semiconductor industry—United States—History. 3. Electronic industries—United States—History. 4. Electronics—Materials—United States—History. I. O'Connell, Sheila, 1968- II. Title.
  HD9696.S44M467 2009
  338.7'62138150973—dc22
                                         2009019379

Printed in the United States of America at Walsworth Publishing Company

TABLE OF CONTENTS

	Acknowledgments	6
Letter from President and Chief Executive Officer Ahmad Chatila		7
Chapter One	*The Birth of Technology*	8
Chapter Two	*MEMC's Pioneering Years*	14
Chapter Three	*A Second Generation of Technology*	26
Chapter Four	*Great Strides in Technology*	38
Chapter Five	*A Global Company*	50
Chapter Six	*Strictly Business*	64
Chapter Seven	*Equations for Success*	72

ACKNOWLEDGMENTS

The editor acknowledges the following people as invaluable in helping design, write, and publish this book. Thank you for your time, effort, and belief in this project.

Marsha Adkins

Hidenor Aihara

Linda Apprill

Dr. KiMan Bae

Albert Chen

Peter Chen

Karla Chaney

Cheryl Collins

Dr. Bruce Kellerman

Donna Kemm

Brad Kohn

Dr. Harold Korb

Dr. DongMyun Lee

Frank McLaughlin

Joy Messner

Bill Michalek

FROM PRESIDENT AND CEO
Ahmad Chatila

From its modest beginnings as a small part of a division of a chemical company in Missouri to its position today as a major global enterprise, MEMC can look back on a long history of success.

When we broke ground on our first facility to produce wafers for the semiconductor industry in 1959, no one could have predicted that the market for silicon wafers would be greater than 45 billion square inch units a year across two industries. Over the next 50 years MEMC technology would be at the heart of breakthroughs that helped shape our modern society. Silicon, and the silicon wafer products that MEMC manufactures, has enabled the modern integrated circuit which enables better healthcare, better communications, and even better entertainment. It has also advanced the burgeoning solar energy industry, by enabling us to efficiently harness clean renewable energy in effort to make solar power a more cost effective alternative to fossil fuels based energy systems. For MEMC this is just the beginning of the journey.

Today, just as 50 years ago, research, development and innovation at MEMC are geared to solving our customers wafer needs of the day, including providing wafer solutions to improve their cost position and product offerings. MEMC currently holds more than 600 worldwide patents on silicon products and processes, plus more than 300 worldwide patent applications currently on file. Our focus on research and development in every area of silicon, and our long-standing tradition of quality have culminated in a company that is better positioned than at any point in our history to meet the demands of the highly competitive electronics and solar industries.

The sacrifice and the true dedication of thousands of employees have enabled MEMC to survive and thrive for 50 years, a rarity in technology companies. We are eternally indebted to them. We are proud of our history, and our innovations, and we know we can achieve so much more. MEMC is only 50, and we know that what we have built will continue to grow. At the same time, our vision for the future encompasses more than silicon wafers for the semiconductor and solar industries. MEMC teammates are constantly seeking out new high growth opportunities, industries and sectors where we can utilize our solid R&D base and expand, both locally, and globally.

I am honored to be leading MEMC in this critical time of our history. The technology made possible through the production of silicon wafers has revolutionized forever how we create, disseminate, retrieve and store information and how we generate energy. The people of MEMC, and the products we develop and manufacture, have made that possible. We are proud of our success over the past 50 years, and look forward to an equally successful future.

Sincerely,

Ahmad R. Chatila
President and CEO
MEMC Electronic Materials, Inc.

CHAPTER ONE
The Birth of Technology

This is the story about a large corporation that, once upon a time, started as a small company and grew into a global force that helps make technology possible. From its inception, the company that is today MEMC Electronic Materials, Inc., was a driving force in the technology that brought us the Information Age.

At no other time in human history has the way we work, the way we play, and the way we communicate with each other changed as rapidly as it has in the period from 1970 to the present. Though the experiments and inventions of Thomas Edison in the late 1800s are considered the birth of the electronic age, it was the continuing research that took place in the first half of the twentieth century that laid the foundation for the Information Age.

> **DID YOU KNOW?**
> On February 6, 1959, a patent application was filed for a "Solid Circuit made of Germanium." Subsequently, Texas Instruments was issued U.S. patent number 3,138,743 for "Miniaturized electronic circuits." In 2000 the importance of the Integrated Circuit (IC) was recognized when Jack Kilby was awarded the Nobel Prize in physics.

When World War II ended in 1945, technology largely remained in the hands of universities, the military, and industry. Postwar life in America was good—relieved of the scarcities of war and full of promise. But even then, homes were still heated largely by coal, which was delivered by truck and dumped through a chute to the basement. Food was stored in an icebox, literally, in an insulated cabinet cooled by a block of ice. Clothes were laundered using a wringer washing machine and then hung on the line to dry. If you were lucky enough to have a telephone, your call was connected by a switchboard operator and subject to eavesdropping by all the other homes that shared your "party line." Most households had a radio, their only source for music, news,

This model was constructed to demonstrate how the earliest transistors worked in 1947.

and broadcast entertainment. For the rare family that owned a television, the few shows that came over the waves were black-and-white and were broadcast for only a few hours each day.

THE TRANSISTOR

The lack of technology began to change when a research team consisting of John Bardeen, Walter Brattain, and William Shockley assembled at Bell Laboratories to find a replacement for the bulky and inefficient vacuum tube that Ambrose Fleming developed in 1904. On June 30, 1948, Bell Laboratories unveiled the transistor, which was smaller, more efficient, more reliable, and cheaper than the vacuum tube. Bardeen and Brattain received a patent for their invention in 1951, and the trio won the 1956 Physics Nobel Prize.

INTEGRATED CIRCUITS

By 1956, transistors were commonplace, being used in everything from radios to the earliest computers. However, the entire electronics industry was looking for a way to overcome the "tyranny of numbers," a phrase used to describe the dilemma that arose from the rising number of wires required to perform increasingly complex functions. In the late 1950s, two men, Jack Kilby at Texas Instruments and Robert Noyce at Fairchild Semiconductor, were working simultaneously but independently to overcome this challenge; their inventions would herald the start of a changing world.

TRAVELING AT THE SPEED OF ELECTRONICS

Thomas Ray, former St. Peters' manager of Quality Assurance from 1981 to 1994, offered the following comparison of the transportation and electronics industry, taken from a very early training program used at the St. Peters plant.

"If transportation (going back to the horse and buggy days) had advanced as fast as electronics over the past twenty years, the results would be truly astonishing. A Bell Laboratory scientist has calculated that, with equivalent advancements, your car might travel at one-quarter million miles an hour, get several hundred thousand miles per gallon of fuel, and be cheaper to throw away than to park at the airport overnight."

The first integrated circuit.

1959	One wafer, one chip.
1969	A dime-sized area of silicon held almost 10,000 transistors.
1985	A dime-sized area of silicon held several hundred thousand transistors.
1990s	Millions of transistors resided on a silicon chip.
2001	Circuitry on silicon can be shrunk to 0.13 microns—1/800th the width of a human hair—and the fastest chips operate at more than two billion calculations per second!

Both men are credited with the invention of "the chip," the monolithic (meaning formed from a single crystal) integrated circuit. While Jack Kilby successfully used germanium as the semiconductor in his model, it was Robert Noyce's silicon version, and the fabrication techniques he devised, that made the integrated circuit practical. Both Texas Instruments and Fairchild Semiconductor filed for patents in 1959, and a legal battle ensued that lasted for over a decade. Patent No. 3,138,743 was issued to Jack Kilby and Texas Instruments for miniaturized electronic circuits in 1964. Patent No. 2,981,877 was granted to Robert Noyce for the silicon-based integrated circuit. Eventually, Texas Instruments and Fairchild Semiconductor agreed to cross-license their technologies. Integrated circuits (ICs) gave device manufacturers the ability to integrate large numbers of transistors, performing multiple functions, in a small area. The "chips" were mass-produced, using a photolithographic process, making them much cheaper and more reliable than the transistors that were constructed one at a time.

EARLY SILICON PRODUCTION

Prior to 1960, the standard for microelectronics manufacturing was referred to as "grown junction" production technology. This meant that the semiconductor device was actually grown as a part of the process of producing single crystal silicon. This end-use device growth is why companies like Westinghouse, IBM, and Siemens were among the earliest silicon producers. The development of planar semiconductor device technology and the development of a reproducible Czochralski silicon production process essentially divorced the device technology from the silicon industry. Dr. Ramo Pelling, manager of the first commercial semiconductor-

Bardeen, Brattain, and Shockley (seated) on the cover of *electronics* magazine September 1948 "Crystal Triode" issue.

grade silicon production facility at DuPont and later the first president of Monsanto Silicon, observed, "As the industry started to grow in the 1960s, companies like Monsanto with their chemical materials background caught up to the semiconductor manufacturers quickly (in terms of technology)."

MEMC'S MONSANTO ROOTS

MEMC Electronic Materials, Inc., originated as part of the Inorganic Chemical Division of the Monsanto Chemical Corporation. Founded in 1901 by John F. Queeny, Monsanto started out manufacturing the artificial sweetener, saccharin. Over the years,

Dr. Wallene Derby, head of the Dayton, Ohio, research lab, which later relocated to St. Louis, Missouri.

Float Zone reactors for polysilicon production at Merano, 1962. Montecatini's Merano plant began experimental production of silicon as early as 1958.

Single crystal production at the Merano plant, 1962. Montecatini went through several transformations, eventually being acquired by Dynamit Nobel in 1980, the company that would later become an integral part of MEMC Electronic Materials, Inc.

The Novara plant in the 1970s. Montecatini merged with Edison in 1964 to become Montedison. Montedison then established a new company, SMIEL, in 1974, which built the plant in Novara in 1976 for its wafering operations. In 1980, both the Merano and Novara plants became part of Dynamit Nobel Silicon.

Monsanto grew and acquired other companies. One of particular note was the Thomas & Hochwalt Laboratories in Dayton, Ohio, which Monsanto purchased in 1936. It was in Dayton that several of MEMC's pioneers, Drs. Robert Walsh and Wallene Derby among them, got their start performing research on gallium arsenide for the manufacture of light emitting diodes (LED). Eventually, the Dayton lab relocated to Monsanto's world headquarters in St. Louis, Missouri.

In 1959, Monsanto broke ground at the St. Peters plant site cementing its commitment to the silicon wafer industry. The developments in silicon technology that took place at the St. Peters plant in the coming years would form the foundation for a technological explosion that continues to this day.

Sources:
http://www.pbs.org/transistor/album1/index.html
http://www.xnumber.com/xnumber/kilby.htm
http://www.computerhistory.org
www.semi.org/en/P043595

MOORE'S LAW

Gordon Moore, for whom Moore's Law is named, started his career at Shockley Labs in 1956 and went on to co-found Fairchild Semiconductor in 1957 and Intel Corporation in 1968. Widely known for his 1965 prediction that the number of transistors on a computer chip would double approximately every year, he refers to this prediction as "a wild extrapolation of very little data." Though his original prediction was only for the ensuing ten-year period, he reviewed it in 1975 and, based on technological advancements at that time, adjusted it to doubling every twenty-four months and extended it for another ten years. Asked about the future durability of Moore's Law, he said, "We're getting pretty close to molecular dimensions in the devices we're making now, and that's going to become a fundamental limit in how we can continue to shrink things. So, it's going to change after another two or three process generations—I don't know exactly when."

The Birth of Technology

CHAPTER TWO
MEMC's Pioneering Years

> **DID YOU KNOW?**
> The Czochralski process takes its name from Jan Czochralski, a Polish scientist who discovered this method for pulling crystal rods as early as 1916. It is said he discovered the Czochralski method in 1916 when he accidentally dipped his pen into a crucible of molten tin rather than his inkwell. He immediately pulled his pen out to discover that a thin thread of solidified metal was hanging from the nib. The nib was replaced by a capillary, and Czochralski verified that the crystallized metal was a single crystal.

ST. PETERS, MISSOURI

When Monsanto Chemical Company broke ground for the St. Peters plant in 1959, it was the newest addition to their Inorganic Chemicals Division. The plant was to manufacture "ultra-pure silicon metal, a material used in the manufacture of transistors and rectifiers," according to J. L. Christian, the division's general manager. This new venture was the culmination of several years of research conducted at Monsanto's two research labs (St. Louis and Dayton, Ohio) and represented Monsanto's first major step toward supplying products for the emerging electronics industry. St. Peters, approximately thirty miles west of St. Louis, Missouri, was a small rural village in 1959, and the clean air of the countryside was considered ideal for the extreme degree of cleanliness required for silicon wafer production.

The people who stood at the groundbreaking in St. Peters could not imagine the range and scope of the products that would one day be possible as a result of the work accomplished at this plant and at MEMC plants around the world. Few people can provide a better perspective than Stanley T. Myers, president and CEO of Semiconductor

1959 press release.

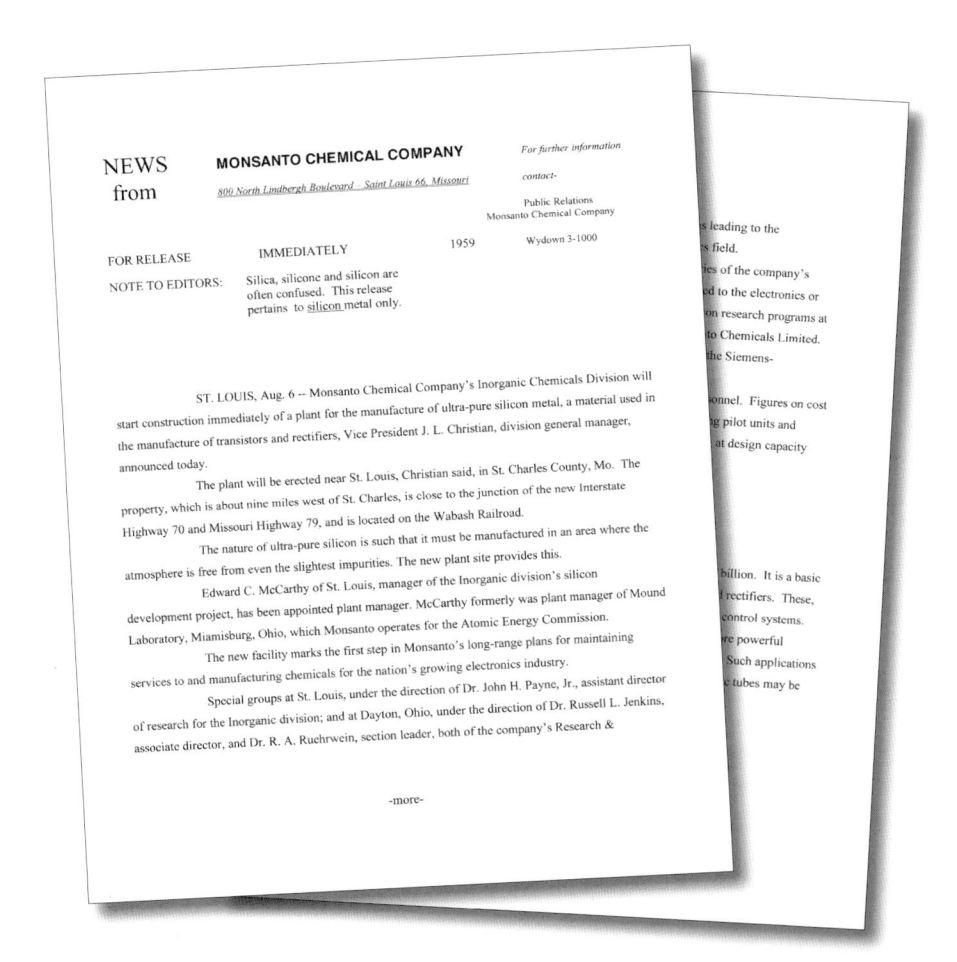

pretty picture." Luckily, he reports, though it did happen a few times, no one was injured. Compared to the extreme safety measures in place today at MEMC, it is difficult to imagine taking such risks. However, risk taking is a quality inherent in pioneers, and pioneers they were. MEMC produced its first three-quarter-inch (19mm) silicon wafer in February 1960.

Polysilicon was produced using the Siemens-Westinghouse process for which Monsanto had purchased a license. Dr. Henry W. Gutsche, who

Equipment and Materials International (SEMI), who worked at the St. Peters plant during its early years. In an interview he gave as part of SEMI's oral history project, Mr. Myers respectfully refers to the pioneering engineers at St. Peters as "a bunch of cowboys" and recalls that the first crystal-growing equipment was all "hand-grown," for the first ten years or so. Sounding just a little astonished at his own recollections, he said, "My first crystal-growing piece of equipment was a converted drill press, if you can imagine. The chamber was made up of a quartz inner tube, and then an outer tube was Pyrex. We circulated water between the quartz and the Pyrex to remove the heat, and the crystal was inside the quartz tube. You wouldn't do that today for anything because it would blow up on you if you got a crack in the glass. If the water hit the crystal, it wasn't a

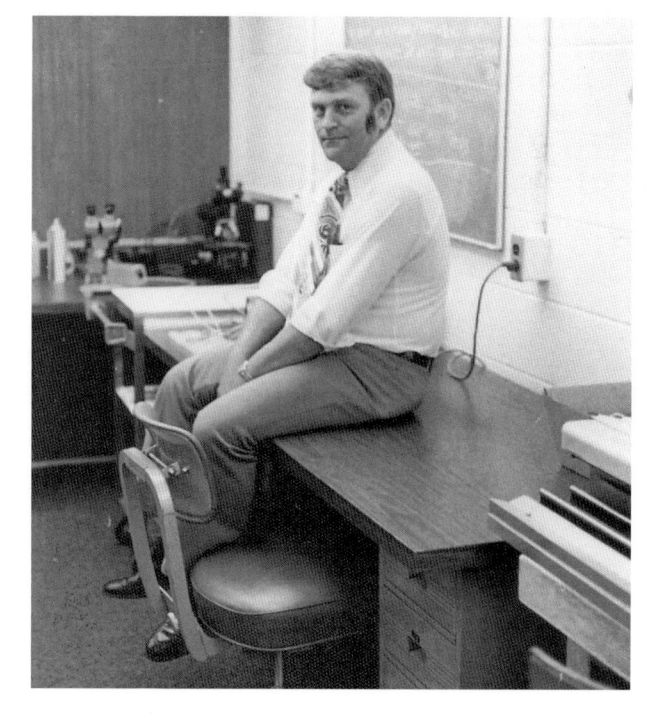

Dr. Stanley T. Myers takes a well-deserved breather.

MEMC's Pioneering Years 15

Timeline

1959
August 6, J. L. Christian, vice president of Monsanto Chemical Company, issues a press releasing announcing construction of a new plant for the manufacture of ultra-pure silicon metal in St. Charles County, Missouri.

1961
Montecatini (Italian chemical company) undertakes the strategy to develop a pilot production of poly and single crystal at Merano. This Italian company would later become part of MEMC.

1962
Dr. Robert Walsh of MEMC pioneers the chemical mechanical polishing process.

The Czochralski (CZ) crystal-growing process is put into manufacturing at MEMC.

1965
Initially, Gordon Moore observes that transistor density is doubling every 12 months. It is later modified to 24 months. The term "Moore's law" was coined around 1970 by Caltech professor, VLSI pioneer, and entrepreneur Carver Mead.

Chemical mechanical polishing of silicon wafers is established at MEMC.

1966
Scientists at MEMC begin developing zero-dislocation crystal.

1968
The first computer with integrated circuits is manufactured.

---

joined the Monsanto Electronics Division in 1964, developed the process that helped make MEMC successful during his earlier years at Siemens.

FLOATING ZONE (FZ) CRYSTALS

The earliest crystals were grown using the Floating Zone (FZ) method and were used in the production of power rectifiers and transistors. Dr. Graham Fisher, current director of Intellectual Property for MEMC, described the process: "It started out with a polysilicon rod, and the process involved creating a molten zone with a moveable heater. The molten zone (floating zone or liquid zone) was thin and was held in place by surface tension. The floating zone could be moved slowly down the rod allowing single crystal silicon to grow behind it, thus converting polysilicon structure to single crystal. This could be repeated multiple times and the crystal would then be very pure, but it was very expensive to do. The crystals were quite small; the wafers were approximately half a millimeter thick, and you'd typically produce about twenty-five to thirty wafers to an inch."

By 1979, Monsanto St. Peters was supplying approximately 80 percent of the U.S. market for float-zoned silicon. The float zone process was phased out in the mid-1980s in favor of another process for producing crystal that had finally matured—the Czochralski (CZ) method.

THE CZOCHRALSKI METHOD

Growing crystals for wafers was job one at MEMC. However, the growing of crystals often meant more capacity, and more space was needed, so the plants would

Early CZ crystal.

Aerial view of St. Peters plant, 1960.

also have to grow. As early as 1962, the original plant building was expanded to provide for the Czochralski method of making single crystal rods. A Polish scientist named Jan Czochralski invented the technique in the early 1900s. Credited for discovering this method of pulling metallic monocrystals, he is still recognized as one of the founding fathers of today's semiconductor technology despite the fact that his research underwent considerable scrutiny in politically torn World War II Europe. In the late 1950s, the process was revisited as a possible method of producing silicon for transistor production and was widely adopted during the 1960s.

The Czochralski method made much larger diameter wafers possible even though it had one disadvantage. Oxygen from the crucible, in which the silicon was melted, contaminated

PIONEERS AND COWBOYS

When you consider how crude some of the early processes of silicon production appear compared to those of today, it's not surprising that words like "pioneer" and "cowboy" come to the minds of those who were there at the time. Some of what they experienced in those early years truly was heroic, and some of it has its comedic side. Consider, for instance, how the first wafers were shipped. While millions of dollars would eventually be spent developing packaging that kept the wafers clean and chip-free, Don Westhoff recalled that the original FZ polished wafers were shipped to a customer (probably GE) in a coin-wrapping package.

MEMC's Pioneering Years

IN HIS OWN WORDS: DR. ROBERT WALSH ON CHEMICAL MECHANICAL POLISHING

"When we started out, there was no integrated circuit industry. All the wafers that Monsanto was selling were just lapped wafers or etched wafers that went into devices—one wafer, one device. They were trying to make these multiple circuits on a wafer and the photographic process they used required very flat surfaces. You can't get that with etching alone. They needed the wafers to be nicely polished, and then epi grown on the surface. The normal polishing techniques used very fine abrasives but no matter how fine an abrasive you use, you damage the surface of the crystal. You couldn't see them in the substrate, but when they grew the layer, they'd show up. Growing an epi layer is a very serious test of how good a surface you have. We couldn't get satisfactory yields and surfaces with the old polishing method. We had to come up with something new.

"The early machines were optical polishing machines. What we found was (we were running with the finest aluminum oxide you could buy), that if you kept polishing with that same suspension of aluminum oxide, if you kept going, eventually you'd get much better surfaces on the wafers. At first we attributed it to wearing down the particle size even further until it was fine enough. Then I realized that what it *really* was, was that we were adding silicon to the polishing solution. These machines would fling the polishing solution off the edge and it would go down and be pumped back up so we were actually adding silicon to the polishing solution. So, thinking about all that, I thought, 'Monsanto makes a product that's a colloidal silicon.' So we tried it one day and it just worked like a charm. It was in water, too, but it was alkaline. We needed it to be alkaline to keep it in suspension. The minute we hit that, the epi yields went up. And as soon as we saw that all the scratches went away on the epi surface, we knew we had something. Basically, that polishing process is what made integrated circuits possible."

Dr. Robert Walsh developed the chemical-mechanical process for polishing wafers and received a North America SEMI Award for his work in 1986.

When asked how it felt to be such a well-known pioneer in the silicon wafer industry, Bob modestly replies, "Well, it's nice," though he received the SEMI Award of North America in 1986, an award established in 1979 "to recognize outstanding technical achievement and meritorious contribution" to the industry. He recalls those early years as "the fun years," saying, "It seemed like every day there was a new challenge to tackle."

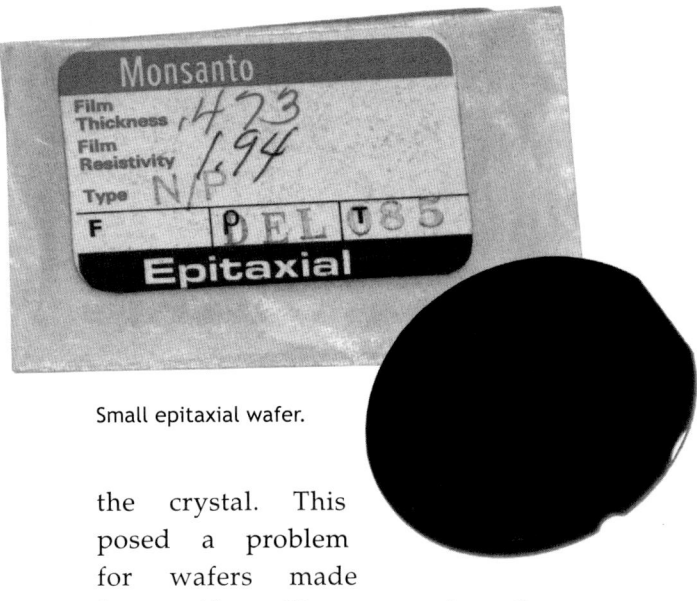

Small epitaxial wafer.

the crystal. This posed a problem for wafers made for rectifiers. However, when the process of gettering was developed, the oxygen was used to attract impurities such as iron, copper, and nickel away from the surface of the wafer, making CZ-grown crystals preferable for semiconductors. Monsanto's work in this area in the 1980s led by Drs. Korb, Craven, and others was very significant. Later work done by Dr. Robert Falster and his team, trademarked in the mid-1990s as "Magic Denuded Zone," cemented MEMC's long history of controlling oxygen behavior in silicon.

MEMC "FIRSTS"

As MEMC marks its fiftieth year of silicon wafer production, the people of MEMC are extremely proud of the "Firsts" this company can claim. Though you will find these "Firsts" scattered throughout MEMC's history, some of the most significant discoveries are attributed to the people who pioneered silicon wafer production at St. Peters.

CHEMICAL MECHANICAL POLISHING

One of the most notable innovations during this time remains an industry standard—the process of chemical mechanical polishing developed by Dr. Robert Walsh.

Men operate BD-2 zone refiners, 1964.

Polishing, October 1970.

MEMC's Pioneering Years 19

RECALLING DR. GUTSCHE

Dr. Robert Sandfort, who was named president and chief operating officer of MEMC Electronic Materials, Inc., in 1989, recalled a few colleagues during those pioneering years at St. Peters. "Monsanto had a first-class technical operation and was able to accommodate itself to all kinds of personalities just to get the expertise that came with it." Among those brilliant, standout personalities was a man named Dr. Henry Gutsche. Dr. Gutsche joined Siemens in 1954 as a research chemist at their laboratories in Pretzfeld, Germany. In 1956, Siemens received a patent for the work Dr. Gutsche did to produce the first semiconductor grade silicon which was 1000 ohm cm P-type uncompensated Si single crystal. In 1964, after a few years with Merck, Dr. Gutsche joined Monsanto Electronics Division in Research and Development and continued his pattern of outstanding achievements. Dr. Sandfort fondly recalled the slouch hat that Dr. Gutsche wore tilted off to one side and remembers him as being "quite a character—a very sentimental man," a trait that Dr. Sandfort attributes to the time Dr. Gutsche and his wife spent in a Nazi prison camp during World War II. In 1979, Dr. Gutsche received a SEMI award for his outstanding contributions to silicon wafer technology. Eccentric—perhaps; a brilliant pioneer—most certainly.

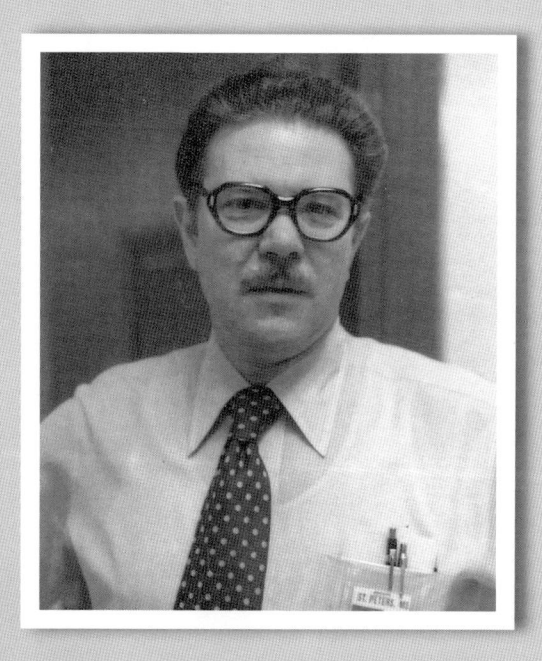

Dr. Henry Gutsche was responsible for setting up the Siemens process at St. Peters.

Up until 1965, Monsanto St. Peters produced and sold unpolished wafers for discrete devices like silicon-controlled rectifiers. With the development of integrated circuits, the device manufacturers required a more perfect wafer surface.

From 1960 to 1965, Dr. Robert Walsh worked at Monsanto's St. Louis research center developing a process to make epitaxial (epi) wafers after having spent several years in Dayton, Ohio, developing a process for single crystal gallium arsenide for III-V semiconductors. When he came to St. Peters, Dr. Walsh took on the task of devising a polishing process that would provide a more perfect wafer surface for growing epi layers.

At the time, wafers were 1" in diameter and 152 microns thick with a tolerance of 19 microns. The very fine abrasives used for polishing the wafers created scratches in the surface. Though they weren't visible in the substrate, they appeared once the epi layer was grown. The ultimate solution that Dr. Walsh devised was the chemical-mechanical method of polishing with colloidal silica in which the mechanical component was used to maintain the flatness of the wafer, and the chemical component kept the surface damage-free. The process of chemical-mechanical polishing is a key enabling factor for production of integrated circuits.

Up to this point, Dr. Walsh recalled, Monsanto hadn't shown any interest in selling polished wafers. "Everybody was trying to make perfect surfaces. When IBM got our wafers, they became very interested because they had their own internal material department and they couldn't make those perfect surfaces," he recalled. In fact, IBM managed to purchase the license for chemical-mechanical polishing, trading some patents with Monsanto as part of the deal. Within the next couple of years, sales of polished wafers greatly increased.

EPITAXIAL LAYERS

By 1964, 1" wafers went into commercial production at St. Peters and shortly thereafter were replaced by 1.5" wafers. Then, in 1966, the first epi reactors were built and installed, giving MEMC the ability to produce a wafer with a superior surface. In epi, a thin layer of silicon is grown on top of the polished slice surface by heating it to a temperature of approximately 2000 degrees Fahrenheit (approximately 1100 Celsius) in a quartz chamber filled with hydrogen gas. Another gas containing silicon in combination with chlorine and hydrogen is introduced into the quartz chamber. The silicon layer grown in this manner is more perfect than the polished slice, having fewer defects and fewer impurities.

The ability to produce wafers with an epitaxial layer increased in significance as device manufacturers required increasingly cleaner and flatter surfaces.

Silicon brochure.

Dr. Graham Fisher described the development of epi as "a platform improvement" saying, "In about 1984, MEMC was the first to commercialize epitaxial layers for CMOS technology. In its way, it was an enabling development to come up with that. People had been putting epitaxial

Distillation columns at the St. Peters expansion, 1970.

layers on silicon for a while; what we did was commercialize it as a blanket layer." The epi layer and the developed MEMC methods gained industry acceptance in the early 1980s as led by Dr. Jon Rossi and his MEMC teammates.

ZERO-DISLOCATION CRYSTALS

In 1966, MEMC began developing zero-dislocation crystals when their best customer at the time, IBM, advised Dr. Horst Kramer and his team that ingots and wafers showing slip patterns would no longer be acceptable. Slip patterns are essentially stress patterns that arise during the process of cooling the crystal, and the term "dislocation" refers to atoms that are out of place in the crystal lattice. Impurities, especially metals, can collect along the structural defects, deteriorating the performance of the device.

Dr. Horst Kramer began his career in silicon crystal growth in 1959 at a company called Knapic Electro-Physics in Mountain View, California. By the mid-1960s, Dr. Kramer was given the task of eliminating slip from pulled crystals. Dr. Kramer's notes record the discovery:

The first experiments showed pretty much what we expected: an increase in pull rate resulted in a more concave interface; all other parameters remaining the same, different seed and crucible rotation combinations could change the interface from convex to concave. The next parameter was the melt level, i.e., the height of the crucible above the melt. In order to save silicon and time, I decided to skip growing the initial melt and start with only half the charge of silicon. The moment we were finished with the neck and started to increase the crystal diameter, we knew we had something special on our hands. The secondary growth lines in the taper were absolutely flat, and reflected light like a perfect mirror. The primary growth lines were strongly ridged and uniform, with no skips. In the body of the crystal, the primaries became flat, and the secondaries were visible only as shadowy lines. The bottom of the crystal, growing in the shape of an upside-down taper, also had pronounced flats, although here they occurred on the primaries. The flats sparkled like the faces of a diamond! We had just grown the first faceted silicon crystal!

As Dr. Kramer noted in his report, his discovery of "Zero-Dislocation" crystal growth "made

Dr. Horst G. Kramer.

the growth of large-diameter crystals possible," saying that he did "not believe that even a 150mm diameter crystal having a low dislocation density could be grown." Further, he explained, "Dislocations are such large crystal defects that they could not be tolerated in today's [1966] device fabrication."

PUTTING DOWN ROOTS

Despite having garnered 80 percent of the market share by the time MEMC opened its doors, business got off to a slow start. Nonetheless, the 1960s remained a period of steady growth

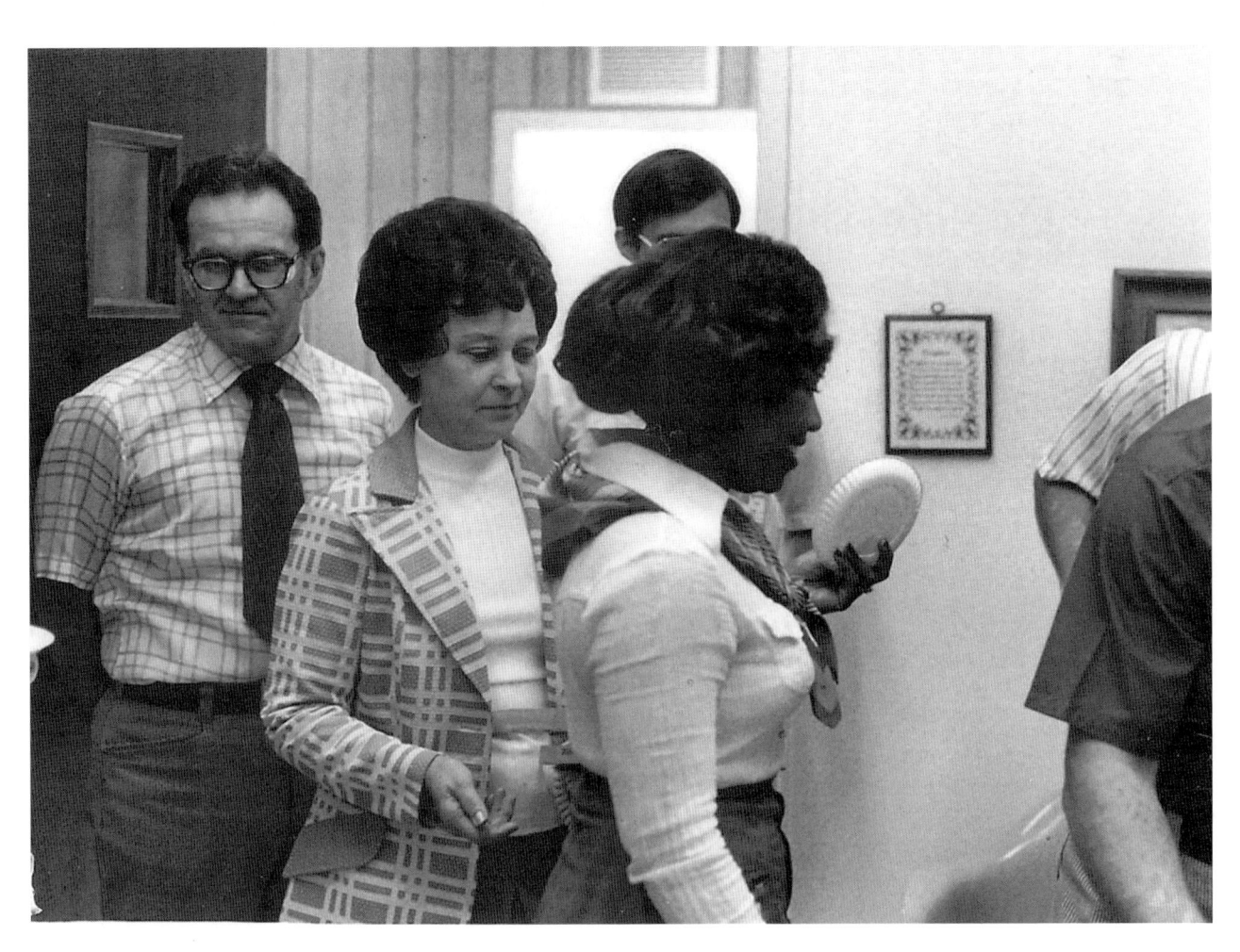

Left is Dr. Bobby D. Stone, who invented St. Peters' zone-refining equipment, BD-1, BD-2, and BD-3.

MEMC's Pioneering Years   23

for the company and laid the foundation for MEMC's future reputation as a leader in the industry. In sharp contrast to its rural setting and the cornfields that surrounded the St. Peters plant, researchers there were giving birth to a technological boom. Monsanto's roots in research and development helped create an environment that bred technological innovations, a trait that remains a hallmark of MEMC Electronic Materials, Inc., to this day.

Additional Sources:
www.semi.org
www.semi.org/en/About/Awards/index.htm
http://www.eoearth.org/article/Czochralski_Jan

Flattening and grinding wafers, 1970.

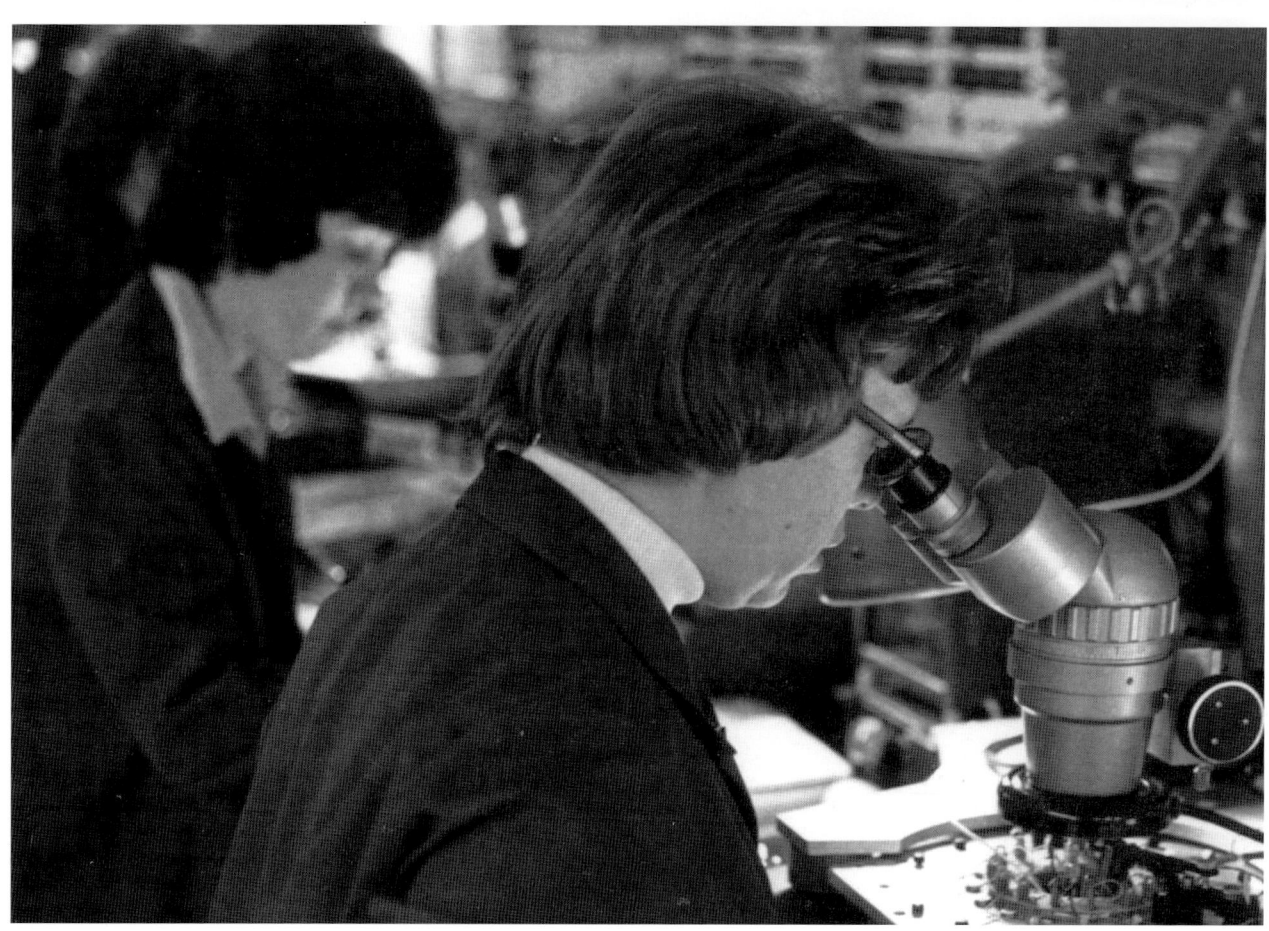

Wafer evaluation, early 1980s.

Left is Dr. Forrest "Frosty" Williams, resident expert in III-V's.

George McLeod was head of Monsanto's electronics business. In the early 1970s, he made the trail-breaking move to go offshore.

Aerial view of St. Peters plant, 1970.

MEMC's Pioneering Years

CHAPTER THREE
A Second Generation of Technology

By 1970, consumer electronics were becoming a growth industry, most visibly in entertainment-related products. The first home video games, played through television receivers, hit the market in 1972. In 1975, CB radios were all the rage along with the "Pong" home video game. The 1980s saw the first personal computers and videocassette recorders coming into the market, and by 1988, CDs were more popular than vinyl records among music enthusiasts.

The 1970s and 1980s were a time of tremendous growth in the silicon wafer industry interrupted by head-spinning highs and lows. In hindsight, the downturns that occurred in the mid-1970s and again in the mid-1980s did so for a variety of reasons. Over-production caused, in part, by optimistic chip manufacturers, competition from the emerging Japanese market, and, of course, the constant drive toward advancements in technology, all contributed to the turbulence of these two decades. Monsanto St. Peters responded to these challenges by focusing on technological excellence combined with constant quality improvement, a strategy that has brought the company to where it is today.

In a study of the fifty-year history of the semiconductor silicon industry, Steven Walsh and others concluded that success in the silicon industry is determined by a company's ability to maintain the highest standards of industry-specific technological competence. He writes, "By 1980, materials companies had overtaken both captive and fabrication and assembly-based producers, and

> **DID YOU KNOW?**
> A man named Ray Tomlinson developed the first electronic mail system in 1971. The first message of any substance was a message announcing the availability of network email and gave instructions to use the 'at' sign to separate the user's name from the host computer name.

ICS AT THE WHEEL

By the early 1970s, integrated circuit technology was largely driving the silicon wafer business. Stanley Myers was Monsanto's Silicon Business Group director in 1972. He recalled, "In the early 1960s we knew [the development of the integrated circuit] would have a significant impact, not only on growing the crystal, but also polishing the wafer—slicing, polishing, and modifying the crystal and putting the epi layers on the wafer. All of these things became much more prominent as soon as IC

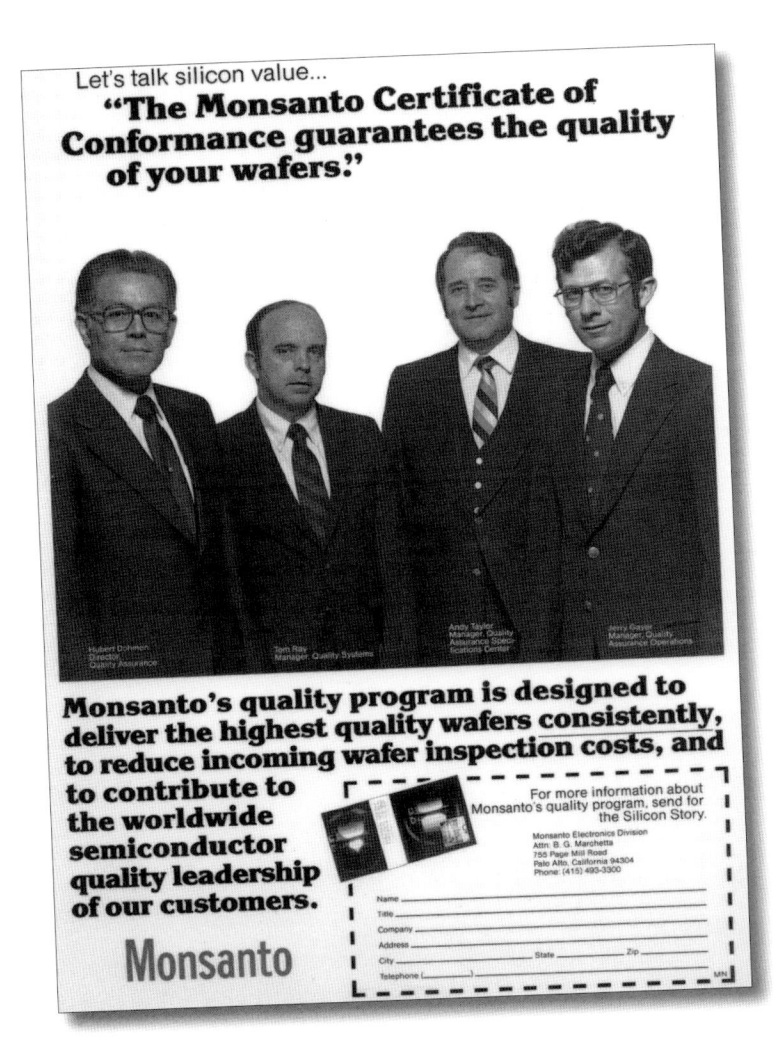

Monsanto advertisement, featuring Hubert Dohmen, Tom Ray, Andy Taylor, and Jerry Gayer.

U.S.-based merchant chemical suppliers dominated the single crystal silicon as well as polysilicon markets. Monsanto was the leading silicon supplier in the world and dominated the worldwide single crystal silicon production with facilities in the U.S., Europe and Asia." According to Walsh, the silicon manufacturers most likely to succeed were those who excelled at both disruptive and evolutionary technology. Disruptive technologies were those innovations that remained proprietary and therefore captive for a period of time; evolutionary technologies referred to the constant incremental improvements on existing processes. Monsanto St. Peters, with its strong background in research and development, was ideally suited for both.

SEMI

Semiconductor Equipment and Materials Institute (SEMI) was formed in 1970 to provide visibility and support to the equipment and materials manufacturers that were the backbone of the burgeoning electronics industry. Having increased marketplace awareness through its expositions, SEMI worked to establish their standards program through which they continue to foster open markets, constructive competition, and cost containment throughout the semiconductor industry. In 1979, they established the SEMI award, given each year to individuals for outstanding technical achievements. When the industry reached global status in the mid-1980s, SEMI redefined itself to stay in step. The "I" in SEMI was changed from "Institute" to "International." SEMI offers its members technical conferences, educational events, and market data collection and analysis and advocates for the industry in public policy, environmental, health and safety issues, workforce development, and investor relations. With eleven offices in manufacturing regions around the world, SEMI serves the manufacturing supply chains for the microelectronic, display, and photovoltaic industries.

A Second Generation of Technology

Timeline

1970
The SEMI organization is founded by Bill Hugle, Fred Kulicke, and John Dannelly.

MEMC builds a plant in Kuala Lumpur, Malaysia.

1971
Electronic mail (E-mail) is invented.

The computer floppy disc is invented.

The microprocessor is invented. It is considered a computer on a chip.

1972
The word processor is invented.

1973
Three-inch wafers go into commercial production at the St. Peters plant.

1975
Four-inch wafers go into commercial production at the St. Peters plant.

MEMC opens Kuala Lumpur, Malaysia, plant.

1976
Montedison builds a second plant in Novara, Italy. This plant would later become merged with MEMC.

1979
MEMC is the first to commercially produce 125mm wafers.

MEMC is the first wafer supplier to control oxygen.

The inaugural SEMI Awards are held, with eleven individuals recognized for their achievement in advancing equipment and materials technology.

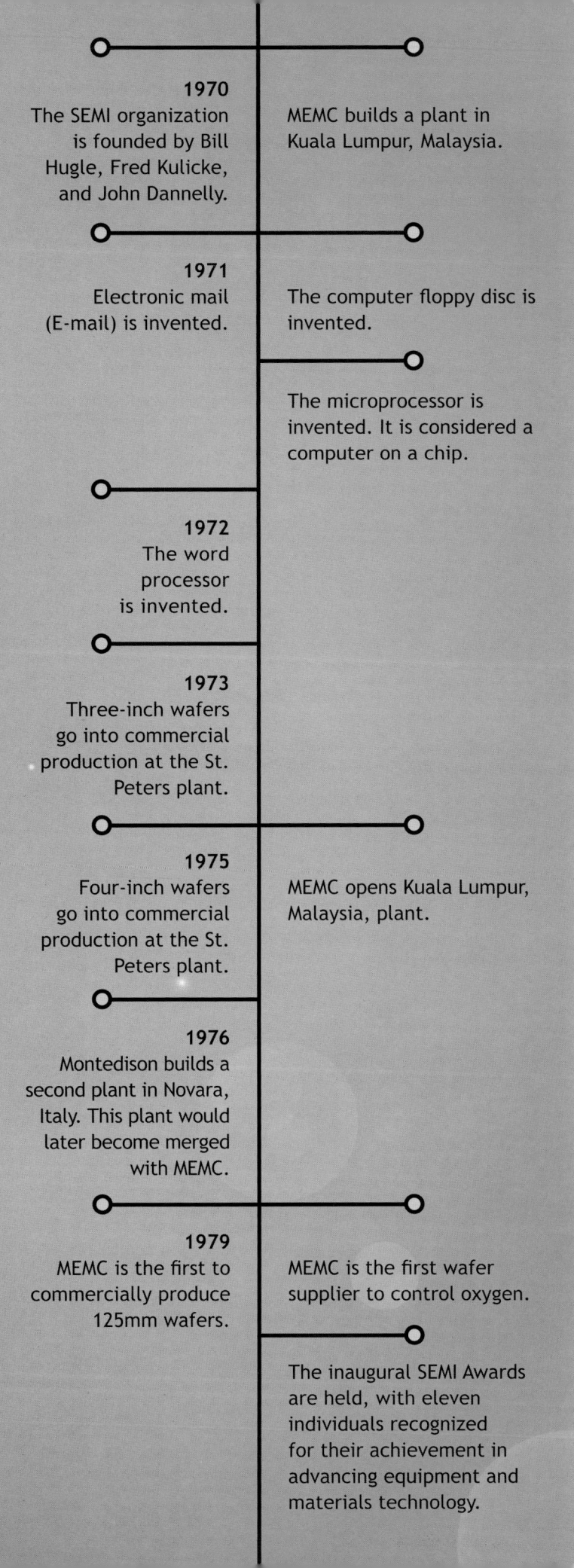

applications took hold and the market began to develop."

Industry experts agree that device manufacturers drove the change toward larger wafers, primarily as a means for cutting costs. Circuits were printed on the wafers using photolithography, and, simply put, the more devices per wafer, the lower the cost. But they also agree that the silicon wafer manufacturers, including MEMC, complied with the demands in part as a means of asserting their technological capabilities. MEMC's customers in the fast-moving electronics industry were putting more and more circuitry on smaller chip areas. To do that, they needed smaller line widths, which called for flatter, cleaner silicon wafers with improved physical and electrical characteristics.

The increased wafer diameters that resulted in lower costs for the device manufacturers came at quite a price for the silicon wafer suppliers, MEMC among them. Each increase in wafer diameter meant an outlay of capital for the equipment required to produce that larger wafer, as well as increased costs incurred to achieve the tighter specifications that were inherent in every diameter increase. As with their customers, cost cutting became a priority for the silicon wafer manufacturers.

MOVING OVERSEAS

The St. Peters plant was expanded in 1962, 1967, 1970, and again in 1974, by which time the plant had doubled from its original capacity. By the early 1970s, competition from the Japanese silicon manufacturers was getting everyone's attention. Stan Myers recalls that in the mid-1970s Japan's silicon production was "probably two generations behind the U.S.," but that "they were on a very, very rapid learning curve."

Kuala Lumpur plant in Malaysia.

MEMC's move toward international status was prompted by competition from the Japanese. The manufacturing site in Kuala Lumpur, Malaysia, built in 1970, was MEMC's first location in the Asian semiconductor market. The Kuala Lumpur site was chosen because of its proximity to the largely English-speaking population of Singapore and because Malaysia's free industrial trade zone provided attractive tax incentives for foreign investors. The most significant benefit, however, was the savings that resulted from lower costs on everything from the physical plant to supplies to labor. The St. Peters plant grew 2.25" crystals, shipped them to the Kuala Lumpur plant for slicing and polishing, and then on to customers.

III-VS

In the 1970s, the St. Peters plant started manufacturing the III-V compounds used to make light-emitting diodes, those red and green lights that first showed up in pocket calculators and digital wristwatches. This operation was eventually dismantled and sold to General Instruments in 1979.

A Second Generation of Technology

Wafer lapping, 1975.

A CHANGING MARKET

In the April 23, 1990, issue of the *St. Louis Business Journal*, Roger McDaniel summed up the Asian influence on the silicon wafer market: "In 1980, half of the silicon sold in the world was consumed in the United States and only 25 percent was consumed in Japan. By 1985, the numbers were reversed as Japan gained dominance in the semiconductor memory market. Now the Koreans and Taiwanese are challenging the Japanese, and the United States and Europe are fighting to stay in the race."

The Milton Keynes plant in England opened operations in 1986. Etched wafers received from the United States were polished and cleaned for Monsanto's European customers, among them well-known names like Siemens, SGS, Thompson CSF, Philips, Telefunken, National Semiconductor, and Motorola. The site included a research and development center and an applications research lab.

SIZE MATTERS

MEMC's focus on providing customers with the cost-saving larger diameter wafers combined with the highest-quality features, led to their reputation as a leader in the silicon wafer industry during these turbulent years and beyond.

In 1975, Monsanto St. Peters was the first to commercially produce 4" (100mm) wafers and again the first to commercially produce 125mm wafers in 1979. The 150mm wafer was first produced in 1981, and at the time, a leading IC manufacturer projected a five-fold increase in chip yield when switching from 100mm to 150mm wafers. In addition, the flatness specification of the 150mm wafer varied less than two microns within a 20mm field of view. In 1984, MEMC partnered with IBM to produce the 200mm wafer.

EVOLVING TECHNOLOGY

In 1975, MEMC became the first wafer supplier to control oxygen in a growing crystal, a process that simply wasn't crucial in the early days. Dr. Graham Fisher explains the significance of the process: "Because the crucible is made of quartz, which is silicon dioxide, in the process of growing silicon, oxygen from the crucible can contaminate the crystal. In the early days, the level of oxygen wasn't very important. As time went on, customers started using more complex processes for making integrated circuits during which the wafer is heated and cooled multiple times. During the manufacture of integrated circuits, the successive heating and cooling causes the supersaturated oxygen atoms to precipitate in the form of small oxide clusters. The higher the oxygen concentration the higher the level of precipitation, so control of oxygen became an important development."

Ironically, solving one problem created another problem in its place because the low oxygen wafers were not as strong mechanically. Graham recalls

Monsanto's MEMC tradeshow booth.

the time spent in the early 1980s trying to get the right balance of oxygen and understanding how the wafer would behave in a customer's device line. "We were trying to get the oxygen at just the right level so when the customer would process the wafer on the fab line, the oxygen would start to precipitate but only to a certain degree and not near the wafer surface," he explains. "This way you maintain the mechanical strength, but the precipitates form in the bulk and not near the surface where the device is formed. An even bigger advantage of this distributed precipitation behavior is that the precipitates trap any unwanted metal impurities that might diffuse into the wafer in the bulk region away from the active device region at the surface. The technique is known as 'intrinsic gettering.'"

About the same time, MEMC added edge grinding to the basic wafer process. Starting with the 100mm wafers, MEMC began grinding the edges to a rounded profile. Before this step was added, the crisp, squared edge of the wafer was more susceptible to chipping. Ground edges meant less waste, and therefore reduced costs, for both MEMC and their customers.

100mm wafers and ingot.

A Second Generation of Technology 31

Left to right: Dr. Dave Keune, Dr. Bob Craven, and Dr. Harold Korb in the Applications Research laboratory, 1975.

Zone refiner.

1ST OZONATED DE-IONIZED WATER FOR WAFER CLEANING

In 1983, MEMC achieved three very significant "firsts" in the area of silicon wafer cleaning with ozone. They were the first to use ozone in a silicon wafer cleaning process, first to use the combination of HF and ozone in a silicon wafer cleaning sequence, and the first to design and install cleaning equipment capable of performing an ozone cleaning process step for silicon wafers.

These innovations were driven by MEMC's recognition that the traditional cleaning chemistries, referred to as Standard Cleans 1 & 2, could not achieve the surface metals levels that were going to be required by the industry, given the purity of cleaning chemicals available. At that time, MEMC was simultaneously developing new methods for surface aluminum analysis that supported the work in cleaning. When surface metals, especially aluminum, were too high, the customer observed non-uniform oxide growth rates as a result.

Employees wear Clean Silicon Group t-shirts that were part of an advertising campaign to establish Monsanto's wafers among the cleanest on the market.

Ingot slicing.

Rack cleaner.

SPARTANBURG LEARNING DISCOVERY

In 1980, MEMC built a plant in Spartanburg, South Carolina, and began commercial production of 6" (150mm) wafers. When production began, the plant had ten thousand square feet of space and forty-five employees. Combined with the Malaysia plant and expansions at St. Peters, Spartanburg brought MEMC's total manufacturing capacity to 330 million square inches per year, a 400 percent increase over that of 1978. John Kauffmann, currently senior vice president of Worldwide Sales, Customer Service, and Marketing, started his career with Monsanto at the Spartanburg plant and recalls that there was a worldwide learning process that took place between plants. "At our plant we made an etched wafer so it wasn't a completely finished product and we were shipping it to our plant in Japan. At that time, I was a technology leader and these wafers that we thought were really, really good opened up a learning exchange between how we did things in Spartanburg and how things were done in Japan." It was such an all-encompassing learning experience that it led to a task force, which John spearheaded, to help improve the quality of the product coming out of the United States.

Dr. Larry Shive, MEMC Senior Fellow, commented, "This work was far ahead of any previous or subsequent cleaning and was truly revolutionary." The process was developed by Wilbur Krusell and co-workers in 1982 and was first installed in St. Peters as the ozone rack cleaner in 1983. It was then copied at MEMC factories around the world. Ozone cleaning technology is now a key cleaning step in every wafer and device factory in the world, giving device manufacturers a predictable and uniform oxide growth rate.

Dr. Larry Shive.

MULTISTRATE SILICON WAFERS

A 1984 brochure entitled "Monsanto: Technology in Silicon" touts the advantages of their MultiStrate silicon wafers, a new class of multilayer products. By this time, device manufacturers were building thousands of ICs on a single chip, requiring improved circuit performance and yield. As a result, the electronic circuitry interacted more dynamically with the physical properties of the silicon. As the brochure said, "the cutting edge of semiconductor technology is now within the silicon."

MultiStrate silicon wafers were a new class of integrated silicon substrate products. The MultiStrate product line was a family of new epitaxial wafers for applications involving high-density silicon-gate CMOS and NMOS. Trademarked EpiStrate, this series of products

John Kauffmann, senior vice president of Worldwide Sales, Customer Service, and Marketing.

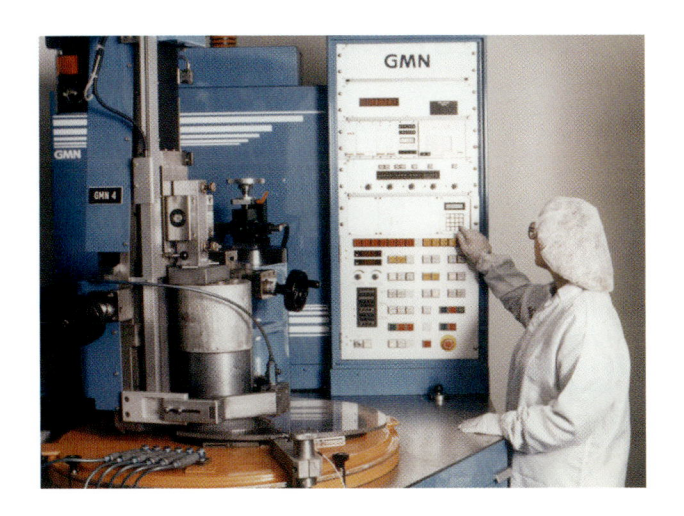

Inner diameter saw with back grinding capability, which reduced the amount of required lapping.

STC horizontal saw, late 1970s.

combined the superior flatness of the basic Czochralski wafer with an epitaxial layer grown on the polished surface. With the MultiStrate product line, substrate characteristics could be matched to the epitaxial process using a Monsanto proprietary technique for engineering the epitaxial/substrate interface to enhance large-scale circuit performance and yield.

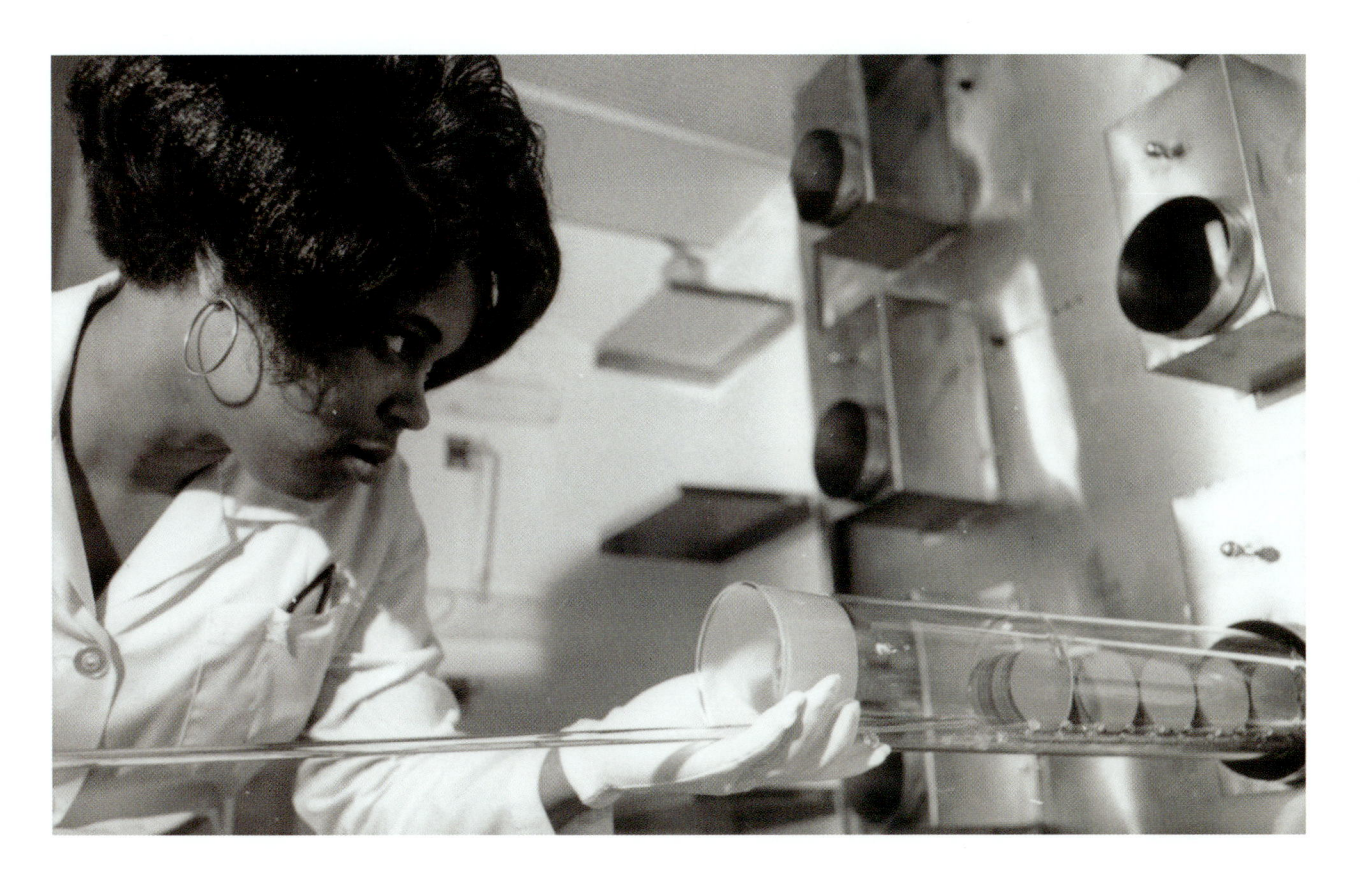

Wafers are inserted into and extracted from a warming oven, slowly allowing them to heat or cool to avoid dislocations in the crystal lattice.

A Second Generation of Technology 35

APPLICATION-SPECIFIC WAFERS

The introduction of the MultiStrate product line perhaps marked the beginning of the application-specific approach to increasing customer demands. In the late 1970s and early 1980s, Monsanto was instrumental in SEMI's efforts toward standardization of silicon wafer production. This was important, primarily because the more standardization Monsanto could incorporate into its manufacturing process, the easier and more cost effective it was to accommodate the customers' requests for specific features as device manufacturers required increasing customization.

Mortin Sentman slicing wafers.

Polishing two-inch wafers.

The highly competitive nature of the consumer electronics products industry created a cost consciousness that trickled down to the silicon wafer industry. The challenge for the wafer manufacturers was to accommodate the device manufacturers' demands for increased customization while simultaneously maintaining their own cost competitiveness. Perhaps the key to MEMC's survival and success was its good fortune at being birthed by a large company with the deep pockets required to stay on the cutting edge of a technology that everyone believed would someday pay off. When MEMC was featured in the September 1990 issue of the *St. Charles County Business Network*, the article read, "Since the early 1960s the giant silicon company has had to keep up with an extremely demanding and competitive market where [companies] who fail to keep up with advanced technology are out of the ballgame in short order." Nearly twenty years later, MEMC's vast technological experience and highly competitive business model has, in turn, made it extremely difficult for new competitors to enter the market.

Additional sources:
"The semiconductor silicon industry roadmap: Epochs drive by the dynamics between disruptive technologies and core competencies," Steven T. Walsh et al., Elsevier, Inc., 2003.
www.semi.org
http://openmap.bbn.com/~tomlinso/ray/firstemailframe.html

MEMC's plant in Spartanburg, South Carolina, was built in 1979-80.

CHAPTER FOUR
Great Strides in Technology

In 1980, with the basic technology of silicon wafers firmly established and the sophistication of integrated circuits increasing rapidly, the very nature of the silicon wafer industry changed. From 1950 to 1980, progress in the semiconductor industry was driven primarily by materials science. That shifted in 1980 as process sciences began to dominate silicon wafer production as customers began to demand application-specific wafers.

In response, MEMC amplified its focus on quality, technological innovation, and customer satisfaction. Application engineers were assigned to customers to better understand their silicon wafer needs. At times, the customers' processes were modeled in-house to ensure the performance of MEMC's products in the customer's wafer fabrication lines. Simultaneously, MEMC's key strategy was to constantly improve its own manufacturing efficiency to maintain its technological superiority and cost competitiveness.

DID YOU KNOW?
In silicon wafer manufacturing, an impurity level of one part per billion equates to one foot in the distance to the moon.

1ST GRANULAR POLYSILICON

In 1991, MEMC's Pasadena facility was the first company in the industry to develop a process using granular polysilicon. In the original chunk polysilicon process obtained from Siemens, rods of polysilicon were produced by the decomposition of an ultra-pure gas containing silicon, hydrogen, and chlorine. The rods were crushed into manageable chunks, which were then placed into the CZ puller crucible by hand. Silicon is relatively hard, and the crushing process posed a purity problem since the hammers that were used to crush the silicon also contaminated it. The chunks of polysilicon had to be hand-stacked by human beings; therefore, the CZ puller had to be at room temperature for charge stacking. After loading the crucible with charge, the CZ pull chamber was closed, purged and evacuated, baked-out and heated to melt the silicon.

In the late 1980s a new polysilicon process to produce granular polysilicon was developed at what is now known as the MEMC Pasadena plant. This process produces beads of silicon, approximately 1mm to 2mm in diameter, with the same or better level of purity. Having small beads of polysilicon eliminated the need for crushing and all the problems and extra steps that arose with it. The beads could be poured into the crucible like a liquid, which meant the crucible could be filled much more easily, automatically, and without the need for human contact and the risk of contamination that resulted. Also, the puller could be filled with polysilicon much faster, with the puller atmosphere purged and relatively hot, thus shortening the turn-around time for the puller and increasing its production capacity.

Inspired by the container for Pringles potato chips, Monsanto invented the Flip Tran Cassette container in 1981 to ship wafers with dry nitrogen to ensure cleanliness. Wafers were canned in a controlled atmosphere to keep the wafer surface pristine.

Granular polysilicon is poured into a quartz crucible.

MEMC ad touts the benefits of granular polysilicon.

Great Strides in Technology   39

1979
Sony introduces the Walkman.

MEMC begins construction on its Spartanburg, South Carolina, plant, which is completed in 1980.

1981
MEMC is the first to commercially produce 150mm wafers.

MEMC ships wafers in the "Flip-Tran" cassette. The "can" is sealed with dry nitrogen to ensure cleanliness.

1982
Technical center is introduced in Edogawa-ku, Tokyo.

MEMC pioneers epi wafers for CMOS applications.

1983
Programmer Jaron Lanier first coins the term "virtual reality."

1984
MEMC is the first to commercialize 200mm wafers in partnership with IBM.

1985
MEMC is the first wafer manufacturer to develop polysilicon gettering.

Japan overtakes the U.S. as the world's leading semiconductor producer.

Microsoft invents Windows program.

Chunk polysilicon is melted.

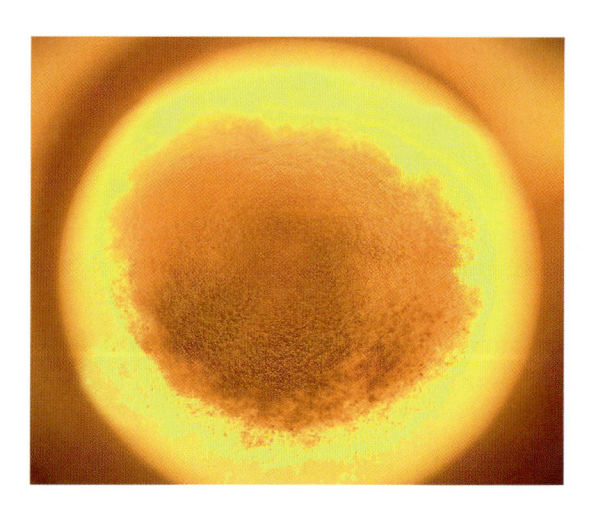

Granular polysilicon is melted.

1ST POLYSILICON GETTERING

As semiconductor devices shrank in size, they became more susceptible to deleterious effects of metallic contamination present in the semiconductor-device-fabrication sequence. Besides degrading the performance of the device, trace metals can form surface silicides that ruin devices by disrupting the device fabrication steps such as etching and dopant diffusion.

MEMC was the first to commercialize polysilicon gettering, a technique for trapping the minute quantities of metallic contaminants present in silicon device processing away from the areas where the devices are to be built. The technology, developed in 1985, involved both internal gettering and external gettering.

In external gettering, a fine-grain layer of polysilicon is grown on the backside of the wafer to draw impurities away from the critical regions of the wafer where devices are built. Wafers with this backside EG layer yield better and are less sensitive to process upsets that might ordinarily degrade or destroy devices. Monsanto (MEMC) was the first wafer manufacturer to offer external gettering, a process developed under the lead of Dr. Dale Hill. By the late 1980s, almost 50 percent of all wafers made by Monsanto had the polysilicon gettering layer applied to them.

Another external gettering process adopted by MEMC was called custom-designed gettering or CDG. Sandblasting of the wafer backside can create damage points at the back surface. These damage points would also effectively trap metallic impurities. By changing the sandblasting conditions to control the density of damage points, a wide range of damage levels could be created in order to provide the correct gettering effect to optimize the customer's device yields.

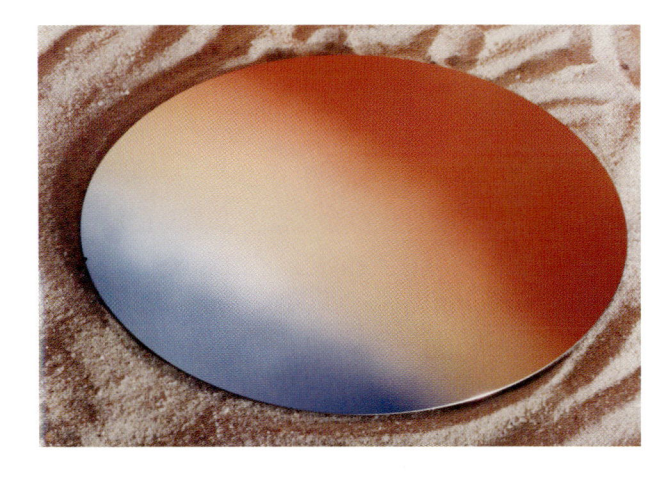

MEMC 300mm wafer.

MEMC's process of internal gettering is featured in the November 2003 issue of *Solid State Technology*.

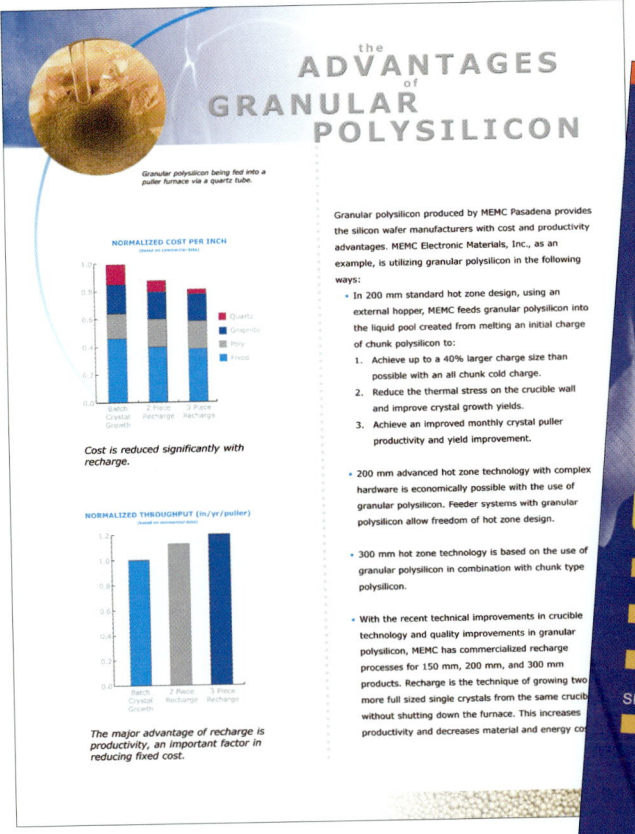

A product brochure describing the benefits of granular polysilicon as produced at the MEMC Pasadena plant.

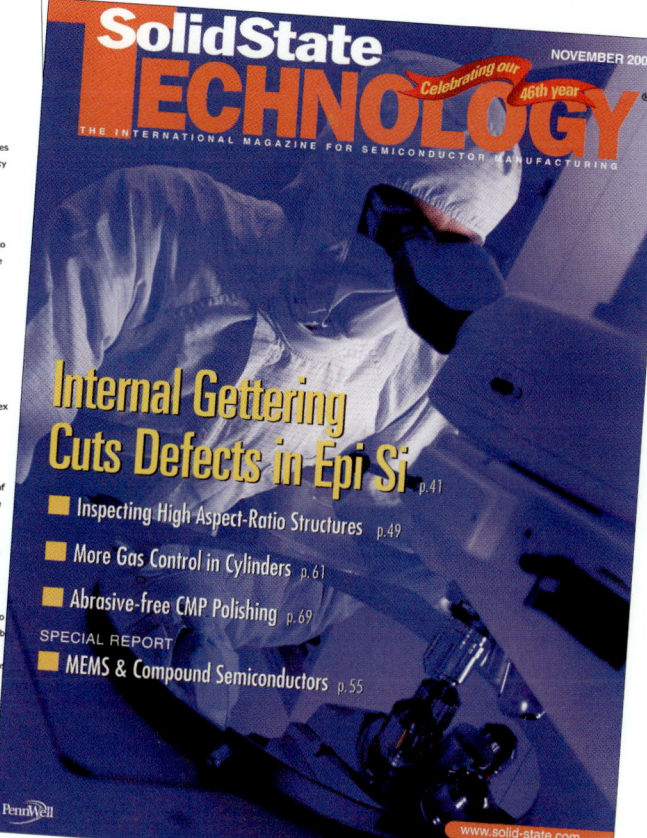

Great Strides in Technology   41

1986
MEMC builds the Utsunomiya, Japan, plant—becomes the first non-Japanese wafer maker with manufacturing and research facilities in Japan.

SEMI establishes the U.S. Japan Trader Partners Conference to ease trade friction between the U.S. and Japanese semiconductor industry.

1987
SEMI and the Semiconductor Equipment Association of Japan enter into an agreement to exchange market data on a monthly basis.

1988
Digital cellular phones invented.

1989
High-definition television invented.

MEMC Japanese teammates band together to build quality products for customers.

1ST UTSUNOMIYA, JAPAN

In 1986, the opening of the Utsunomiya plant made MEMC the first non-Japanese wafer producer to have manufacturing and research facilities in Japan. Unlike the move to Kuala Lumpur, which was largely motivated by opportunities for cutting costs, the Utsunomiya plant reflected an emerging strategy for locating close to the customer. Among the Japanese customers, Nippon Electric Company (currently known as NEC) was the largest and, acknowledging Monsanto as a world leader in silicon wafer production, sought them out as a back-door-to-front-door supplier. A short-lived attempt at warehousing proved to have little benefit to MEMC; silicon wafers do not respond well to warehousing, and NEC's specifications for wafers were constantly changing.

Despite these early challenges, the 300mm Center of Excellence was added to the Utsunomiya plant in 1998, and it remains an integral part of MEMC's presence in the Asian and Japanese markets.

MEMC JAPAN

Originally built in 1986, the Utsunomiya plant is located approximately one hundred kilometers north of Tokyo where MEMC Japan is headquartered. In 1997, the Utsunomiya plant underwent a $250 million expansion for the production of 300mm wafers and currently produces 200mm and 300mm polished silicon wafers and 200mm and 300mm epitaxial wafers. With the support of the Tokyo sales office, the Utsunomiya plant is dedicated to the highly competitive and advanced Japanese and Asian markets.

Utsunomiya is the capital and most populous city of Tochigi Prefecture, Japan. Historically speaking, the city is known for the Battle of Utsunomiya Castle, a battle between pro-Imperial and Tokugawa shogunate forces during the Boshin War in May 1868. Modern-day Utsunomiya, however, is renowned for its jazz and gyoza.

The Utsunomiya, Japan, plant opened in 1986.

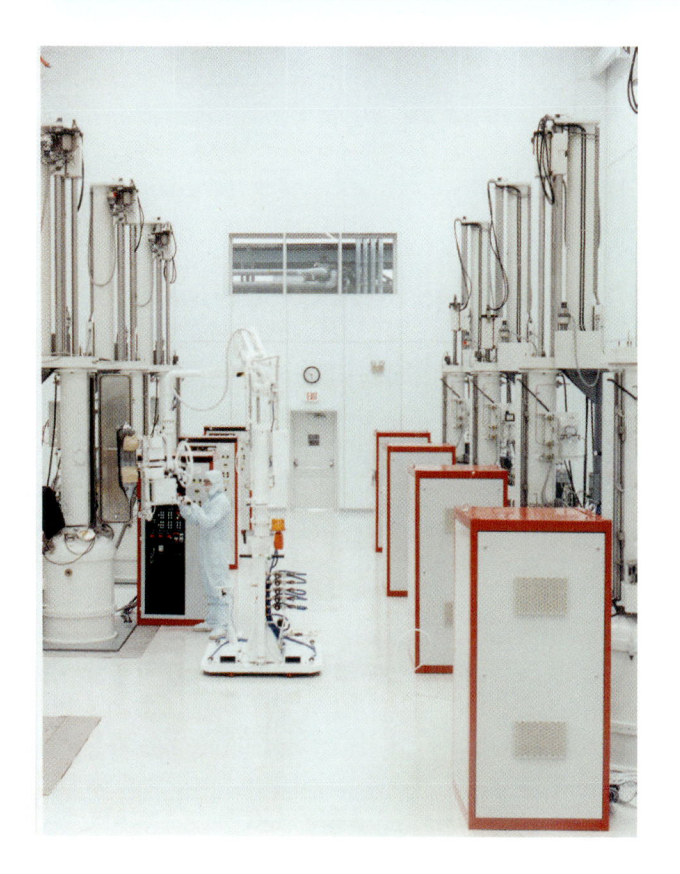

1ST 200MM WAFERS

In 1984, MEMC became the first company to commercialize 200mm wafers, a feat achieved in partnership with IBM. As with every previous increase in diameter size, the demand for quality rose and specifications increased. In 1990, MEMC announced the purchase of several million dollars worth of equipment used to make 200mm diameter wafers from IBM. Ralph Hartung, vice president for U.S. manufacturing at the time, called the 8" wafer "a whole new generation that is becoming a boom business." The purchase of equipment from IBM allowed MEMC to supplement existing production capacity to help the company maintain its position in the silicon wafer market.

200mm crystal production at St. Peters.

Great Strides in Technology

Ralph Hartung, vice president, US Manufacturing. Photo from the early 1980s.

Liz Dulick leads a tour of post polishing through final cleaning at the St. Peters plant.

DRAM, SRAM, AND EPROM

DRAM or Dynamic Random Access Memory is a type of random access memory that stores each bit of data in a separate capacitor within an integrated circuit. Random access means that locations in the memory can be written to or read from in any order, regardless of the memory location that was last accessed.

SRAM or Static Random Access Memory is faster and less power hungry than DRAM. Owing to a more complex internal structure, SRAM is less dense than DRAM and is therefore not used for high-capacity, low-cost applications such as the main memory in personal computers.

An EPROM, or Erasable Programmable Read-Only Memory, is a type of memory chip that retains its data when its power supply is switched off.

An article in the October 1992 edition of *Semiconductor International* explained the significance of the 200mm diameter wafer to the semiconductor industry:

Semiconductor manufacturers are gearing up to use 200mm wafers primarily for high volume, large area circuits such as DRAMs, SRAMs, flash-EPROMs and HDTV related memory. Officials at MEMC Electronic Materials foresee, over the next five years, the sale of 200mm wafers will be increasing by 800 percent.

CONTINUED INNOVATIONS

Up until 1992, the Inner Diameter Saw, or ID saw, was the standard operation for slicing wafers. For the production of 200mm wafers, the move was made toward wire sawing. With wire sawing, one complete ingot is placed in the saw, and a single helical wire makes multiple cuts through the ingot at the same time. Though much faster and more efficient than the ID saw, the potential downside is that if something goes wrong you risk losing the entire ingot. One advantage, however, is that while the diamond ID saw cut was about 350 microns wide, the wire saw is 200 microns wide, creating less waste per cut and more wafers per ingot.

Also notable is the fact that, during this time, the controls of the crystal-growing furnaces were changed to computer-digital systems, giving the operators finer gradations of control. It was one of those strange twists of fate where the silicon wafer MEMC was making was improved through the use of the silicon-based circuits in the computerized control system.

HEADLINE: "MONSANTO TO SELL SILICON WAFER FIRM"

Although the front-page article in the November 14, 1988, issue of *Electronic News* announcing Monsanto's plans to sell its electronic products division didn't exactly read like an obituary, for some it felt like one. As the nation's sole producer of 8" silicon wafers, news of the impending sale sent shockwaves through Silicon Valley, and many feared it was an indication that the United States was losing its technological edge. Worse, the Department of Defense feared that the United States' security could be jeopardized should silicon wafer production, a critical defense technology, leave the country.

While some American companies, including IBM, Texas Instruments, and Motorola, had their own internal wafer production, none of these facilities was large enough to fill the gap should a crisis of national security occur. One Silicon Valley insider was quoted in *Electronic News* saying, "This is it. It's all over. Monsanto was the last stand."

Headline announces Monsanto's intent to sell MEMC.

Above: Wafer slicer, 2009.

Below: This diagram explains both the inner diameter saw and wire sawing of wafers.

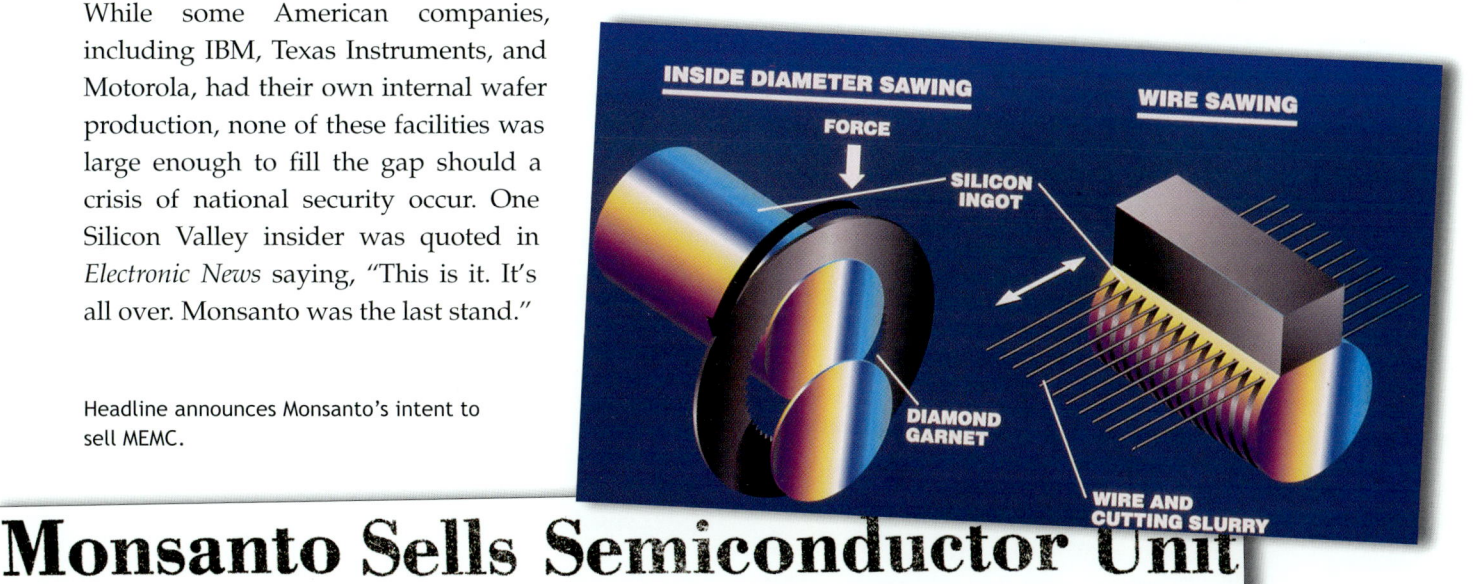

Great Strides in Technology 45

In the past several years, polysilicon has become an important and successful part of MEMC's business, and Merano plays a big part in that success.

Though Monsanto's electronics products division had long been a financial drain to the company, in a strange turn of events, it posted operating profits of $8 million for the first three quarters of 1988 on sales of $137 million. Monsanto had gone through a reorganization just three years prior, and since that time, speculation ran high that the company would renew its allegiance to chemicals, agriculture, and biotechnology products and exit the electronics materials business.

An offer to purchase came from Hüls AG of West Germany, a subsidiary of VEBA AG. The previous year, Hüls announced its intention to purchase Dynamit Nobel Silicon (DNS), a large manufacturer of polyvinyl chloride as well as other petrochemicals, detergents, coatings, and specialty chemicals with sales of $1.9 billion in 1986 for primarily intermediate chemicals, among them trichlorosilane used to make polysilicon.

Dr. Robert Sandfort recalled the concerns surrounding the sale to Hüls: "There were a lot of misgivings. Really, we would have preferred that Monsanto retain ownership of the company, but they did not want to do that. Monsanto was in the position of concentrating its efforts on life sciences, biotechnology. So we saw that they were going to exit the business, and we thought, 'There has to be somebody to buy it.' We would have preferred that a U.S. company step forward and buy it, but they didn't."

Novara, Italy, plant.

DYNAMIT NOBEL SILICON

1957	A company named Montecatini begins research on silicon in Novara, Italy.
1961	Montecatini constructs a pilot plant in Merano, Italy.
1964	Montecatini merges with Edison. The new company is named Montedison.
1974	Montedison establishes a new company, SMIEL, and begins wafer production in 1975.
1976	SMIEL builds a new plant in Novara, Italy. Polysilicon and single crystal remains in Merano; the wafering operations are moved to Novara.
1980	Dynamit Nobel acquires the silicon operation in Italy and Germany and changes its name to Dynamit Nobel Silicon.
1983	DNS establishes a new technology center in Sunnyvale, California, USA.
1984	DNS constructs a new wafer plant in Research Triangle Park, North Carolina. They ship their first lot of wafers in 1986.
1987	DNS ships its first epi wafers out of Novara, Italy.
1988	DNS's silicon operation in Italy and Germany is acquired by Hüls (VEBA) and becomes DNS Electronic Materials.

Above: Epi reactor.

Left: Epitaxial deposition.

> **EMPLOYEE RECEIVES PATENT**
>
> The July 1988 newsletter read as follows:
>
> Art Ackermann, Engineering Specialist at St. Peters, has received a U.S. patent, which he shares with Peter Tremont, retired CED engineer. The two men perfected a type of ozone technology used to make ultra pure deionized (D.I.) water. This unique process was developed in 1983 right here at our St. Peters Site. In this process, ozone is added to D.I. water to kill bacteria particles, which, if present, can cause defects by sticking to the wafer surface. The process has had a dramatic positive impact on our wafer quality and yield. As a result, all of the Monsanto wafer facilities now use this technology.
>
> In 1986 and 1987 Bob Craven, TSD Superintendent, and Art Ackermann marketed the ozone technology to Monsanto customers. At the end of 1987 the technology was licensed to Arrowhead Industrial Water Company for additional worldwide marketing. To date, direct sales, licensing and royalties have grossed over three million dollars for our company. The technology will not be sold to our silicon competitors. Rather, it is marketed to device manufacturers to improve their yield and competitiveness. We believe that all device manufacturers will eventually utilize the Monsanto ozone technology to improve quality and yields. The ozone process is another example of Monsanto's leading edge technology!

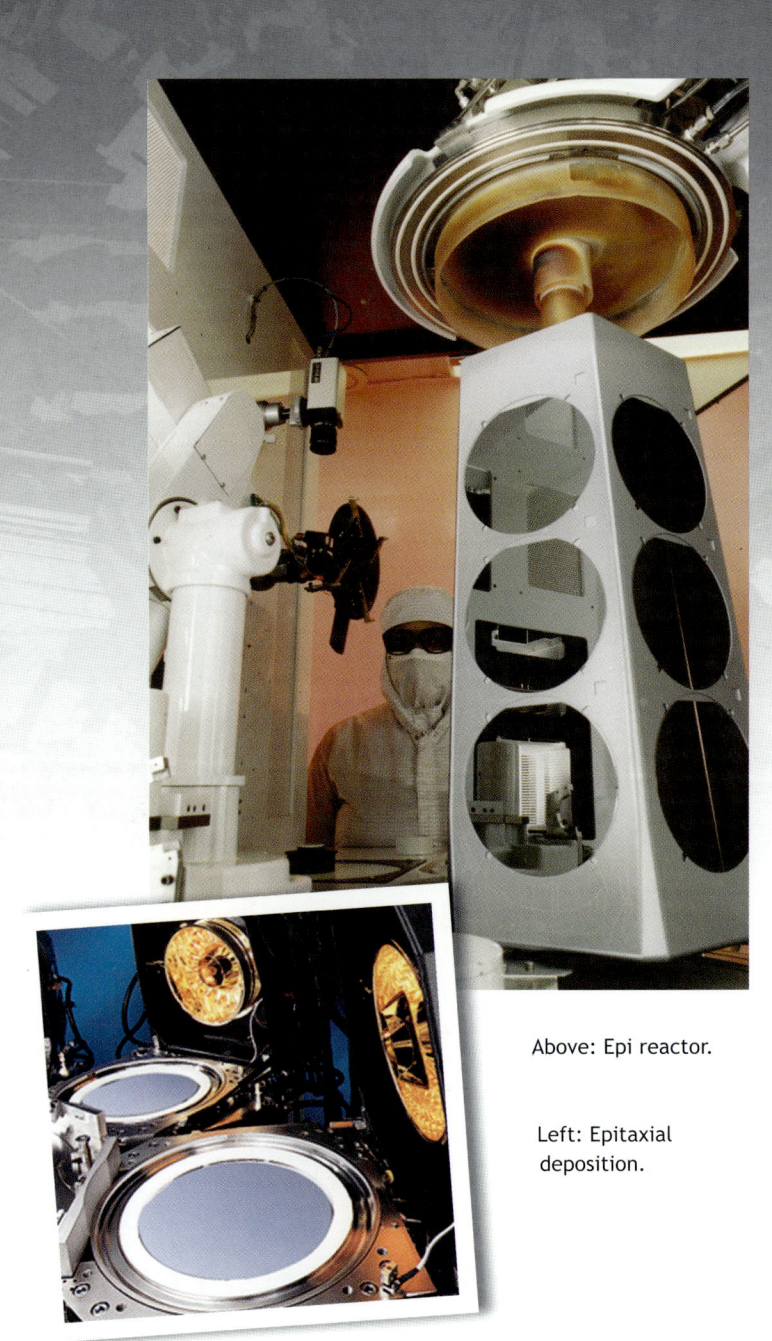

In December of 1988, the federal Committee on Foreign Investment in the United States (CFIUS) announced its intention to review the sale under the provisions of the 1988 Omnibus Trade Act, a law that gave the president the right to block the sale of a United States business to a foreign company if national security was threatened by the sale. Again, Bob Sandfort was in the thick of things: "It [the sale] did raise a lot of concern in Washington, and we had to have a number of meetings there to try and understand their concerns. What they didn't seem to understand in Washington was that we weren't the only people making silicon. The Japanese had two franchises, Shin-Etsu Chemical and Osaka Titanium (currently known as Sumco), and the Germans had two franchises, Wacker and DNS, so

Missouri Governor John Ashcroft with Roger McDaniel, first CEO of MEMC Electronic Materials, Inc., 1989.

they really weren't losing the technology because it was already there."

This was the first time the law was invoked, and despite concerns expressed by representatives of the departments of Commerce and Defense, the Committee on Foreign Investment in the United States (CFIUS) voted unanimously to support the sale. Two primary factors contributed to this decision. First, Hüls assured CFIUS that they were committed to keeping the silicon wafer business in the United States. Second was the announcement that the company would be managed by a combination of current executives as well as Hüls officials.

On February 7, 1989, then-president George H. Bush announced that he had given his permission for the sale of Monsanto Electronic Materials Company to Hüls AG. The West German government had already granted approval for the sale. All that remained to finish negotiations was approval from the governments of Japan and Malaysia where Monsanto Electronic Materials Company had manufacturing sites. Within a couple months of President Bush's approval, the sale was finalized. Shortly thereafter, Dongbu Industrial Company took full ownership of an earlier joint venture between themselves and Monsanto located in the Republic of Korea, named Korsil. Hüls, who already owned the DNS facilities in Merano and Novara, Italy, prudently closed down wafer-making plants in Research Triangle Park, North Carolina, and Milton Keynes, England, in May of 1989.

The new company, MEMC Electronic Materials, Inc., was now the second largest manufacturer of silicon wafers in the world. With Dr. Roger McDaniel named as president and CEO, MEMC launched an entirely new era in the company's history.

MEMC made plans to build a 20,000-square-foot addition to the St. Peters plant to house its new world headquarters. When then Missouri governor John Ashcroft visited the plant, he commented, "The combination of Monsanto and DNS makes MEMC a major player in the world market for this highly specialized product and makes Missouri a world leader in this business."

In the ensuing years following the sale to Hüls, MEMC entered into some strategic alliances and joint ventures, both in the United States and abroad, to enhance their position in the world silicon market.

Sources:
St. Louis Post Dispatch, May 2, 1990
St. Louis Post Dispatch, June 25, 1990
St. Louis Business Journal, April 23, 1990
Semiconductor International, October 1992
http://en.wikipedia.org/wiki/Utsunomiya

CHAPTER FIVE
A Global Company

Within a year of its merger with Dynamit Nobel Silicon (DNS) in 1989, the St. Peters plant, now the World Headquarters of MEMC Electronic Materials, Inc., added twenty thousand square feet to its facility. Capital funding totaling nearly $40 million was approved for the manufacturing and research facility to ensure MEMC's survival in the highly competitive silicon wafer business. Roger McDaniel, then president and CEO, reported that MEMC had about 23 percent of the world market of $1.5 billion for silicon wafers and anticipated sales of $390 million in 1990, an increase of approximately $40 million over 1989 sales. By 1991, MEMC's sales were well over $400 million, and by 1994, the St. Peters plant produced two thousand different wafer types, 90 percent of which did not exist three years prior. In 1995, *Industry Week* magazine named MEMC St. Peters one of the ten best manufacturing plants in the United States.

DID YOU KNOW?
On Earth, silicon is the second most common element, exceeded only by oxygen.

From the time of Hüls' purchase of MEMC to the late 1990s, the company once again saw a surge of technological innovations and industry alliances. MEMC's strategy for secure growth was to expand manufacturing capacity for advanced products, maintain technology leadership in the industry, and focus on strategic alliances with customers and peer companies throughout the industry.

JOINT VENTURE—KOREA

In 1990, MEMC announced its plans to enter into a joint venture agreement with Pohang Iron and Steel Ltd. (POSCO) and Samsung Electronics Company Ltd. to open a plant in South Korea under the name Posco Hüls Co., Ltd. "We have more than half of the Korean market now," Roger McDaniel said. "We need to protect that. Korea is going after Japan in a strong way in the making of memory chips. From the Korean point of

MEMC'S FIRST CEO, ROGER MCDANIEL

In 1996, Roger McDaniel retired following a thirty-four-year career that took him from entry-level chemical engineer at a small plant in West Virginia to the chief executive officer of the second-largest silicon wafer manufacturer in the world. Mr. McDaniel joined Monsanto's silicon business as director of Facilities and Planning in 1980. In 1982, he became the first person with responsibility for the silicon wafer business outside the United States. When interviewed for the Summer 1996 issue of *MEMC Today*, he recalled those years when the industry shifted to international status. "In 1982, our sales were around $100 million, and only around $20 million of that was from international sales. Today, our group sales for just one quarter are equal to the total sales for all of the first four years I was with the silicon business. And, 60 percent of it now comes from sales outside the United States."

When Monsanto sold its electronics products business to Hüls in 1989, McDaniel was in charge of operations. With the formation of MEMC Electronic Materials Company, Inc., he was named president and CEO. Upon his retirement, he commented on the satisfaction he felt watching the company grow. He said, "This organization has a real sense of purpose today. We are good at what we do, and are improving every day."

view, they will have invested $1 billion in this effort, and they will not want to be dependent on a foreign source for their silicon."

To this day, many Koreans consider both Samsung and POSCO symbols of national pride, and their affiliation with MEMC was a boost to the company's international standing. The goal was to put MEMC in a position to compete in any market that developed anywhere in the world with the exception of Soviet-block countries where U.S. restrictions still applied to the sale of high technology products. The manufacturing facility was built at Cheonan City in the south Chungcheong province of South Korea. By 1993, the plant was fully operational for the manufacture of polished wafers.

In June of 1996, a conference was held in Tao Yuan, Taiwan, to discuss MEMC's plans for addressing the growing Asian silicon wafer market, which had more than tripled in the preceding five years. At the time, it was forecast to account for more than a third of the worldwide

300mm wafer.

A Global Company   51

1990
Tim Berners-Lee creates the World Wide Web/Internet protocol (HTTP) and WWW language (HTML).

1991
Eight-inch wafers go into production at MEMC's joint venture in Cheonan, South Korea.

1993
The global semiconductor equipment market exceeds $10 billion in sales for the first time.

The Pentium processor is produced.

1994
MEMC's joint venture, Taisil Electronic Materials, is located in Hsinchu, Taiwan.

1995
MEMC Southwest, a joint venture with Texas Instruments, is established in Sherman, Texas, to produce six-inch wafers.

MEMC trades on the New York Stock Exchange as an Initial Public Offering under the stock symbol "WFR."

1996
Stan Myers, former president and CEO of Siltec and one of MEMC's "Pioneers," is appointed president of SEMI.

1997
MEMC St. Peters begins commercial production of 300mm wafers.

1999
MEMC introduces three new products: Optia, Aegis, and MDZ.

MEMC secures base patents on Perfect Silicon-brand wafers.

MEMC Taisil plant 300mm opening ceremony, Hsinchu, Taiwan.

silicon market by the year 2000, and several expansions reflected those projections. The joint venture plant in Cheonan, Korea, completed an expansion that increased its capacity to more than three hundred thousand 200mm wafers per month.

JOINT VENTURE—TAIWAN

In 1994, MEMC entered into a joint venture with China Steel Corporation, China Development Industrial Bank, and Chiao Tung Bank. The company formed as a result of this joint venture was named Taisil Electronic Materials Corporation. The manufacturing facility was located in Hsinchu, a medium-sized city located about an hour south of Taiwan's capital city of Taipei. Referred to as Taiwan's "Silicon Valley," the Hsinchu location gave the Taisil plant back-door access to many of MEMC's Taiwanese customers. Tommy L. Cadwell, MEMC's president of Asian Operations at the time, said that investment in the plant was expected to exceed $150 million. Taisil was Taiwan's first large-diameter advanced silicon wafer producer. Its capacity upon its opening was eighty-five thousand wafers per month, but expansions quickly brought its capacity to its peak volume of three hundred thousand 200mm wafers per month.

EVEN BY 1995 STANDARDS

According to an article in the October 2, 1995, *St. Louis Post Dispatch*,

At some stages (of silicon manufacturing), the air in the process line is five million times cleaner than the atmosphere in St. Louis.

The circuitry on Intel Corporation's Pentium chip is equal in complexity to a street map of the entire United States and fits on a surface about the size of a fingernail.

That same year, 1994, MEMC also purchased a silicon wafer plant in Santa Clara, California. The facility, previously a subsidiary of Kawasaki Steel Corporation, produced 100mm and 150mm silicon wafers and employed about two hundred people. The plant, known as Kawatech, was purchased to increase arsenic crystal capacity.

THE IBM CONNECTION

IBM and MEMC share an interdependent history that dates back to the earliest days of the St. Peters plant. During the late 1980s, the two companies worked together to launch 200mm wafers. This success was beneficial to both companies as it furthered the relationship, and on many occasions IBM would work with MEMC on special development projects, and MEMC could reap the benefits of the relationship both technologically and financially.

Groundbreaking ceremony at Posco Hüls, MEMC's joint venture in Cheonan, South Korea.

The plant in Cheonan, South Korea—MKC.

Taisil Electronic Materials plant in Hsinchu, Taiwan.

A Global Company 53

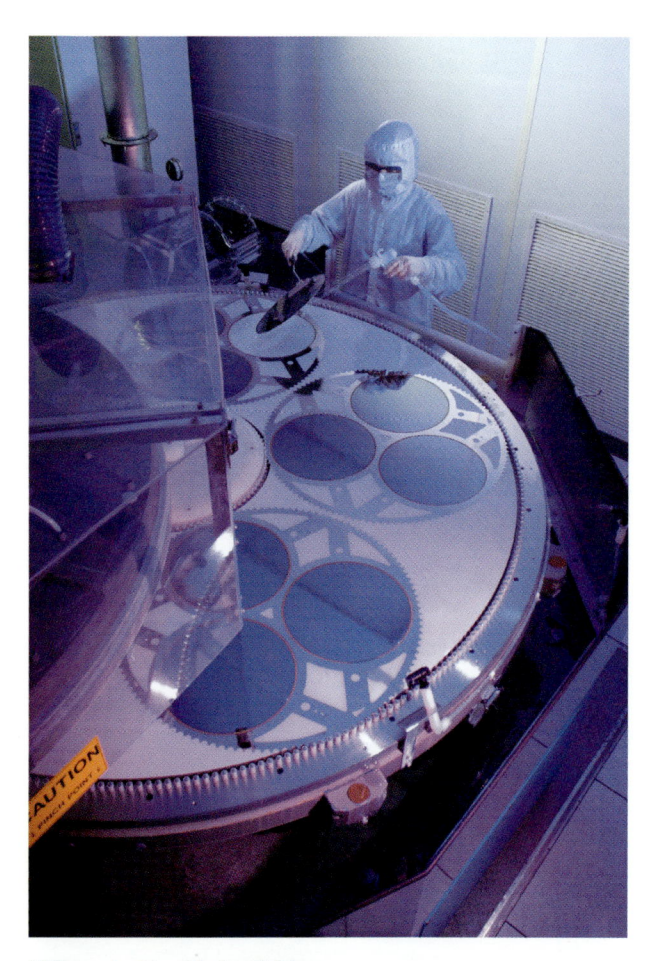

300mm double-sided polishing.

"Silicon-on-Insulator," a.k.a. SOI, wafers were originally manufactured by SiBond, LLC, a joint venture between MEMC and IBM. These wafers were produced in 2007.

PASADENA PLANT

1983	Ethyl completes $150M investment to develop granular polysilicon manufacturing process.
1985	Ethyl completes $70M investment in granular polysilicon manufacturing process—PA21.
1987	Ethyl completes startup of PA21 after investing an additional $50M in the process.
1994	Ethyl spins off chemical business into a separate publicly traded company, Albemarle. Polysilicon manufacturing is part of the new company.
1995	Albemarle sells PA21 to MEMC for $58M.
1998	MEMC completes investment in an additional silane production unit, PA22, raising capacity to 2,500 metric tons per year.
2007	MEMC completes investment in third silane production unit, PA23, raising total capacity to 6,000 metric tons per year.
2008	MEMC completes investment in fourth silane production unit, PA24; this investment combined with expansions in Merano raises total MEMC capacity to 8,000 metric tons per year.

In 1994, MEMC again partnered with IBM to form a new company called SiBond, LLC. The company, based at St. Peters, used technology developed by the IBM Research Division plus manufacturing and sales techniques developed by MEMC in the creation of a specialty technology called "silicon-on-insulator" (SOI) technology to be used for memory devices containing more than one gigabit of data.

The specialty wafer's makeup is a sort of sandwich with a layer of silicon dioxide serving as an insulator

MEMC's Pasadena plant, 2007.

between layers of silicon, providing a substrate for microcircuits that operate on reduced power consumption. The quick response microcircuit worked well in advanced devices such as laptops or hand-held computers and portable telephones. SOI wafers are energy efficient and eliminated the need for heavy batteries previously required for portable devices. SOI technology is currently a major component of some high-speed network services and direct-link entertainment devices.

JOINT VENTURE—PASADENA, TEXAS

Back in 1984, MEMC entered into a joint venture agreement with Ethyl Corporation to develop a process for making granular polysilicon. Ethyl Corporation operated the pilot plant, located in Pasadena, Texas. Then in July of 1995, MEMC purchased the production facility from Albemarle, a company that Ethyl Corporation had spun off its specialty chemical business in years prior, and established MEMC Pasadena. The acquisition of the Albemarle facility in Pasadena shored up the security of MEMC's supply for polysilicon. When polysilicon was in tight supply in the mid-1990s, MEMC was producing more than half of its polysilicon internally and was also able to purchase from Hüls some of the raw material needed to make chunk polysilicon. Strengthening the internal supply chain with granular polysilicon provided MEMC with "a good secret weapon." Located ten miles southeast

A Global Company

MEMC Southwest Groundbreaking Ceremony, October 2, 1995. Left to right: Ron Backschies, human resources manager; J. R. Moore, finance manager; C. B. Lee, QA manager; Bob Bennett, 100mm manager; Jerry Smith, engineering; Bob McCarley, 150mm manager; David Eaton, customer service manager; Frank McLaughlin, 200mm manager; David Fronterhouse, 150mm manager; Mike Grimes, 150mm operations manager; John Robinson, president and COO, MEMC Southwest.

Aerial view of MEMC Southwest, Sherman, Texas, in December 1996.

Crystal puller, first tool installed at MEMC Southwest, November 1996.

of downtown Houston, Texas, MEMC Pasadena continues to produce ultra-pure granular polysilicon, silicon powder, and Monosilane.

JOINT VENTURE—SHERMAN, TEXAS

Texas Instruments started their Sherman, Texas, plant in 1965 for the production of integrated circuits for IBM. In 1966, the Defense Systems and Electronics Group established a fabrication operation in Sherman.

In July of 1995, Texas Instruments issued a press release announcing their intention to form joint venture alliances to share the investment costs caused by rapidly changing and expensive technology, faster product-to-market-cycle time requirements, and rising manufacturing costs. Forming a joint venture made it possible for two companies to work together as partners to enhance production capability while sharing the financial investment.

In October of 1996, the joint venture between MEMC and Texas Instruments was realized at the TI plant in Sherman, Texas. The name given the new company, MEMC Southwest, reflected MEMC's majority interest at the time of the joint venture, though today the facility is 100 percent MEMC-owned. The plant was opened to expand production of 200mm silicon wafers in anticipation of increased demand. At the time, Texas Instruments was making semiconductors and electronic products and, in 1995, was projecting an increase of 30 percent in semiconductor demand. Dr. John Robinson, who was named president and chief operating officer of MEMC Southwest, was quoted in a press release saying, "We are very pleased to begin operations today and immediately supply TI with smaller diameter silicon wafers. We will also begin building for the future by proceeding immediately with plans to construct a new facility for 200mm wafers for both TI and general market demands."

That expansion came to fruition in 1997 and represented an investment of almost $300 million. This freed up MEMC Southwest's existing manufacturing facility for the captive production of TI's smaller diameter wafer requirements.

MEMC, INC.: A GLOBAL COMPANY

Through its joint ventures and other alliances, MEMC implemented a process whereby localized sourcing and technical service provided significant competitive advantages. MEMC's geographically diverse production network allowed the company to service the world's key semiconductor markets while remaining close, strategically and geographically, to its most important asset—the customer. Other than the Korean Plan, which is 20 percent owned by Samsung, currently, all original joint ventures are 100 percent MEMC owned and operated.

Fifty years after its inception as a pioneering plant in St. Peters, Missouri, MEMC Electronic Materials, Inc., remains a global leader in the silicon wafer industry. With a worldwide network for manufacturing polysilicon, wafer production, and finishing, MEMC's scope is truly global.

THE LARGE-SCALE INTEGRATION (LSI) ERA

When first developed, integrated circuits contained only a few transistors. Referred to as small-scale integration or SSI, they used circuits containing transistors numbering in the tens. By the late 1960s, devices contained hundreds of transistors on each chip, called medium-scale integration or MSI. The mid-1970s ushered in the era of large-scale integration or LSI with tens of thousands of transistors per chip. Predictably, the next generation, starting in the 1980s, was that of Very Large-Scale Integration or VLSI, using hundreds of thousands of transistors. The term

Double side wafer grinder at MEMC Southwest's modifications area, April 1997.

First completed crystal at MEMC Southwest, December 1996.

VLSI has since been supplanted by ULSI or Ultra Large-Scale Integration to refer to an integrated circuit with more than one million components per chip.

Each successive generation of IC capacity requires increasing levels of perfection in silicon wafer technology. Throughout the 1990s, MEMC developed silicon wafers that provided optimal solutions for VLSI-era device manufacturers and beyond. The proprietary knowledge they represent

MEMC ADVERTISEMENTS

WANT TO REDUCE YOUR THERMAL BUDGET BY UP TO 40%?
Over 2 million wafers used annually in fabs worldwide.
Avoid capital expenditures by reducing cycle times.
Eliminate high temperature oxygen out-diffusion.
Eliminate low temperature oxygen precipitate nucleation.
Drop-in capability.
A simple solution.
MDZ®
Time is money. Why wait?
MEMC
TECHNOLOGY IS BUILT ON US
www.memc.com

INTERESTED IN YIELD IMPROVEMENTS UP TO 10%?
REACHING NEAR 100% GATE OXIDE INTEGRITY?
SIGNIFICANTLY IMPROVING RELIABILITY?
No "device killing" structural EPI defects.
Eliminate costly out-diffusion, nucleation, and growth thermal cycles.
COMBINE TWO PROVEN SOLUTIONS IN ONE WAFER:
Perfect Silicon™ — silicon that's defect-free throughout the thickness of the entire wafer.
MDZ® — a drop-in solution that provides a deep precipitate-free zone with robust intrinsic gettering.
That's a recipe for surviving below 90nm.
OPTIA™

Technology is built on us...
...for fifty years and counting.
In 1959, MEMC produced its first silicon wafers for the emerging semiconductor industry. Since then, our wafers have enabled significant advances in technology that the world has enjoyed for the last fifty years. At MEMC, we're proud of our past and focused on the future.
MEMC
TECHNOLOGY IS BUILT ON US
www.memc.com
Building on the past. Ready for the future.

LOSING DEVICES DUE TO METAL CONTAMINATION?
Step-function improvement in gettering consistency.
Effectively getter high levels of metals such as Copper, Nickel, and Iron.
Eliminate steps that are likely to add metal contamination.
Demonstrated gettering in next generation low thermal budget processes.
Drop-in capability.
Reliable performance.
MDZ®
Need security?
Get the best insurance policy available.
MEMC
TECHNOLOGY IS BUILT ON US
www.memc.com

DR. ROBERT FALSTER: IN HIS OWN WORDS

In April of 2001, Dr. Robert Falster, Senior Fellow of MEMC Electronic Materials, Inc., received the European SEMI Award for his contribution to the semiconductor manufacturing industry. He described the work that led to this award:

"It was an exciting time. The control and engineering of the precipitation of oxygen in our wafers during our customer's varied manufacturing processes was a huge challenge and, to a large extent, a mystery. There were simply too many variables. This resulted in huge variations in wafer performance. Decades of work on the problem around the world had not brought the silicon world much closer to a robust solution. The solutions at hand were specific to each and every application and resulted often in immense and, ultimately needless, complication and rigidity at many levels of our business. MDZ solved all this at a stroke. Research centered in Novara resulted in our mastery of the engineering of useful vacancy concentration profiles in silicon wafers. We showed that certain depth profiles could be used to act as a template which takes complete control over the wafer's oxygen precipitation behavior and creates an ideal structure for high yielding integrated circuit manufacture. These profiles could be accurately installed in every wafer resulting in completely uniform performance in all our wafers. Immensely important was that we broke the chain of complication that extended from the growing crystal to the finished electronic device. We created a wafer whose behavior was not only ideal but which could be implemented in a way that waas independent of all those interacting components that had bedeviled and tied us up in knots in the past: the details of the crystal growth process, the distribution of oxygen in our wafers and the details of our customer's manufacturing processes. This was a revolution and a relief."

Dr. Robert Falster, 2006.

will continue to ensure MEMC's position in the industry for years to come.

PERFECT SILICON

Perfect Silicon is MEMC's proprietary defect-free crystal growth process designed to completely suppress the formation of low-density, grown-in defects. The main principle behind the CZ single-crystal growing process lies in the rapid transport of growth-incorporated excess intrinsic point defects to harmless sinks before they have a chance to react to form defects. This method of growing CZ single-crystal controls the point defect level by preventing clustering as the crystal cools. The as-grown crystal is free of cluster defects across the wafer, thereby eliminating the need for post-growth engineering such as annealing or epitaxial growth. Perfect Silicon wafers are 100 percent free of crystal-originated particles and direct surface

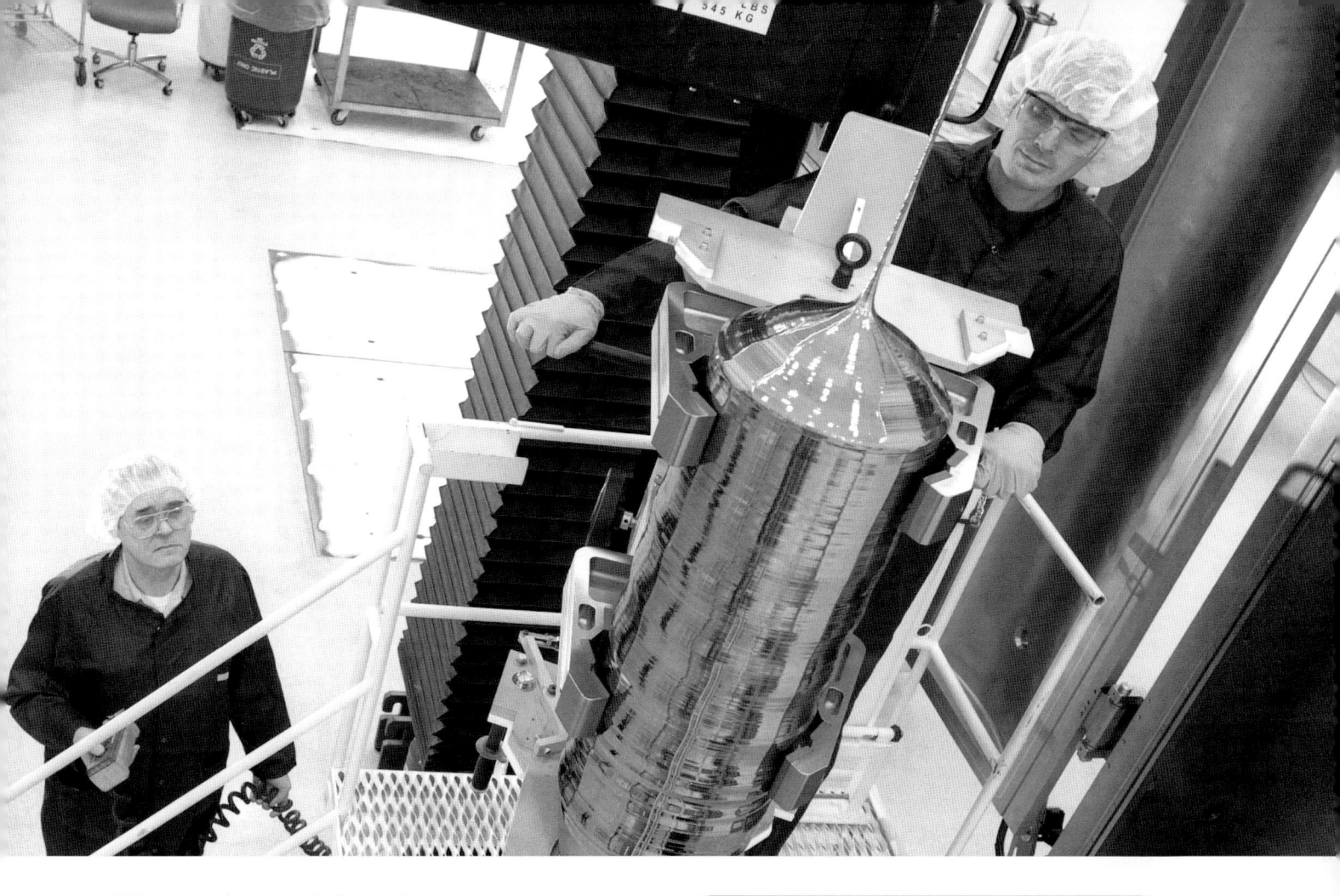

300mm crystal ingot at St. Peters plant.

oxide defects and are the most advanced silicon materials available in the world today.

MDZ

In the mid-1990s, MEMC developed a breakthrough thermal processing technique called the Magic Denuded Zone.

Of considerable concern for IC manufacturers since the VLSI era began was formation of oxygen precipitates in the active device region where they become device killers. Controlling oxygen precipitation during the processing of integrated circuits has proven to be one of the most difficult challenges in silicon wafer defect engineering, especially since they serve a dual good cop/bad cop role. While oxygen precipitates are potential device killers in the near surface, they play an essential role of providing protection against impurities in the bulk by acting as a gettering sponge for impurities such as metals.

> In 1990, the characteristics of a silicon wafer were measured in terms of microns. In the year 2008, wafers are made with a level of purity that permits the equivalent of less than one drop of foreign matter in an Olympic-sized swimming pool and are measured in nanometers (one size greater than an atom), a measure one thousand times more precise than just ten years ago.

Historically, IC fabrication thermal budgets included long hours of oxygen out-diffusion at high temperatures greater than 1000 degrees Celsius followed by oxygen re-nucleation at lower temperature. This would ensure that the near surface region where the device is made is free from

A Global Company 61

OPTIA

In 1999, MEMC was granted the first patents on its MDZ and the Optia product line. Integrated circuits in the VLSI and ULSI era continue to push limits as they strive to maintain Moore's Law with each next generation device becoming increasingly dense with more complexity than the previous architecture. Consequently, increasingly tighter specifications are required for silicon substrates. The optimal solution is to eliminate all potential harmful crystal defects in the surface and bulk. Optia, MEMC's patented premier polished product, was specifically designed not only to meet today's tighter crystal specification but tomorrow's challenges as well. The Optia product line resolves two crystal-related harmful defects at once using MEMC's patented Perfect Silicon crystal growth process combined with MDZ. In Perfect Silicon, crystal voids that lead to device failures are completely eliminated in the crystal growth process, making the wafer completely free of these harmful defects. By applying MDZ, oxygen precipitates in the near surface are eliminated, leaving a deep denuded oxygen precipitates but the underlying bulk of the silicon wafer had sufficient gettering.

Magic Denuded Zone or MDZ is a revolutionary process whereby the oxygen precipitation profile (near surface free depth and bulk density) is controlled in a reliable and robust process that is nearly independent of the initial oxygen concentration or crystal growth history.

The advantage to the IC manufacturer is immediately straightforward. Since MDZ wafers arrive with a robust oxygen precipitation profile template built in, the IC fabrication process required to control these device killers is simplified, saving both time and cost while improving reliability and performance. In short, MDZ saves time, reduces IC fabrication cost, improves reliability, and leads the way for next generation device performance.

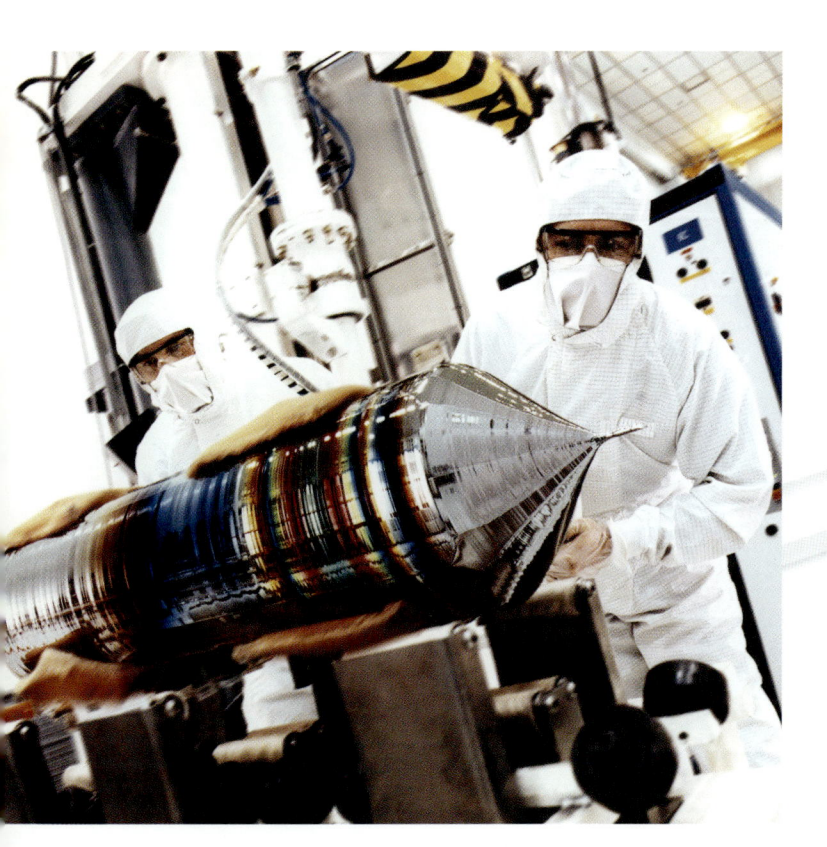

A 300mm ingot is loaded onto a cart for transporting to the next process.

The 300mm wafers in carriers are stocked on shelves in the clean room during processing.

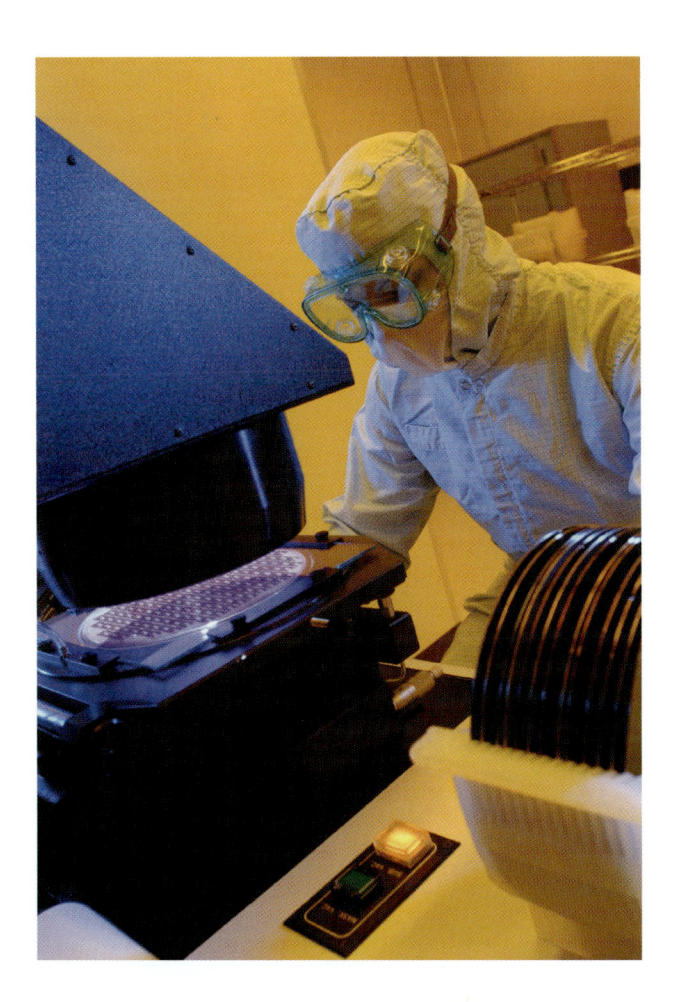

MEMC Worldwide R&D Cleanroom Lab, St. Peters.

zone for IC fabrication. Elimination of all crystal-related harmful defects in the real estate used for IC device fabrication area is an optimal solution for today and future device generations.

300MM WAFERS

In 1991, the push began to develop 300mm wafers. MEMC's customers, working constantly to maintain their own competitive edge, predicted they could achieve about 2.25 times more devices on each wafer with the transition from a 200mm wafer to a 300mm wafer.

While the move to 300mm wafers would entail capital expenditures by the device manufacturers for new equipment and facilities to handle the larger diameter, those costs would not increase in proportion to the surface area, largely because the 300mm wafer meant less wasted silicon near the wafer edge where the rectangular devices meet the curved outer diameter of the wafer.

MEMC began producing and selling 300mm wafers by the end of 1995. The shift from 200mm to 300mm wafers again involved subtle evolutionary refinements of the basic manufacturing process. Tighter flatness specifications were achieved by the development of double-sided polishing, which helped to keep the device lines cleaner. This is because the backside of the wafer was polished smooth and no longer had the rough surface composed of pits from etching. The etch pits acted as particle traps and released the particles during wafer processing, causing some contamination of the device lines. In addition, MEMC added edge polishing to the process. This also helped to keep the device lines clean, especially with the advent of immersion lithography in the fabrications. Though edge grinding was introduced in the manufacture of smaller diameter wafers, edge polishing further increased the wafer's chip resistance.

MEMC was already supplying 300mm wafers to many of the twenty-six leading semiconductor device makers worldwide when the first phase of the 300mm research and development line at St. Peters was completed in March of 1997. The line included a full set of innovative equipment to convert 300mm-diameter silicon crystals into 775-micron thick wafers with ultraflat, polished surfaces.

In October of 1997, MEMC celebrated the opening of St. Peters' state-of-the-art large-diameter wafer production facilities. According to then president and COO, Dr. Robert Sandfort, of the more than $600 million in capital investments spent worldwide in 1996, $150 million of it was spent at the St. Peters site.

Sources:
Electronic News, November 14, 1988
St. Louis Business Journal, April 23, 1990

CHAPTER SIX
Strictly Business

> **DID YOU KNOW?**
> In October of 1995, *Industry Week* magazine selected MEMC's St. Peters plant as one of America's Ten Best Plants based on management practices, employee involvement, quantifiable performance indicators, and evidence of the company's competitiveness.

The year 1995 marked some significant changes at MEMC in what proved to be a banner year for the company. The year prior, the company reported earnings of $34 million in sales of $660.8 million, and by the end of 1995, MEMC celebrated its eleventh consecutive quarter of sold-out production.

GOING PUBLIC

On July 13, 1995, MEMC Electronic Materials, Inc., began trading as a public company under the symbol "WFR" on the New York Stock Exchange. The initial public offering of seventeen million shares of stock raised some $408 million on an offering price of $24 per share. Proceeds from the offering were earmarked for use in reducing MEMC's debt to its parent company, Hüls AG, and to finance the aggressive growth plan that was estimated to cost approximately $700 million through 1997.

RECESSION HITS

In 1996, MEMC had record net sales and net earnings despite the fact that, during the second half of that year, overcapacity and resulting price declines in the semiconductor device industry resulted in reduced orders for silicon wafers. This trend, which continued into 1997, combined with MEMC's significant startup costs for its new and expanded facilities, led to a net loss of $6.7 million for the year.

Despite the gloomy conditions, MEMC continued

Jon Jansky of MEMC St. Peters has the rapt attention of Bob Sandfort, Meyer Gallow, and Ludger Viefhues in 1993.

to position itself for the time when the market would turn around. In March 1997, the first phase of the 300mm research and development line at St. Peters was completed, and 300mm wafers were being shipped to customers worldwide. Additional investments in the Utsunomiya, Japan, facility, totaling some $250 million, were announced in October of 1997.

By 1998, an industry-wide recession was in full swing. It was the first year since 1985 that year-over-year demand for silicon wafers actually decreased. Overcapacity in the semiconductor device industry remained a problem, aggravated by a slowing economy in Asia and out-and-out recession in Japan. Compared to the U.S. dollar, the Japanese yen and Deutsche mark remained strong, further undercutting already weak prices and margins.

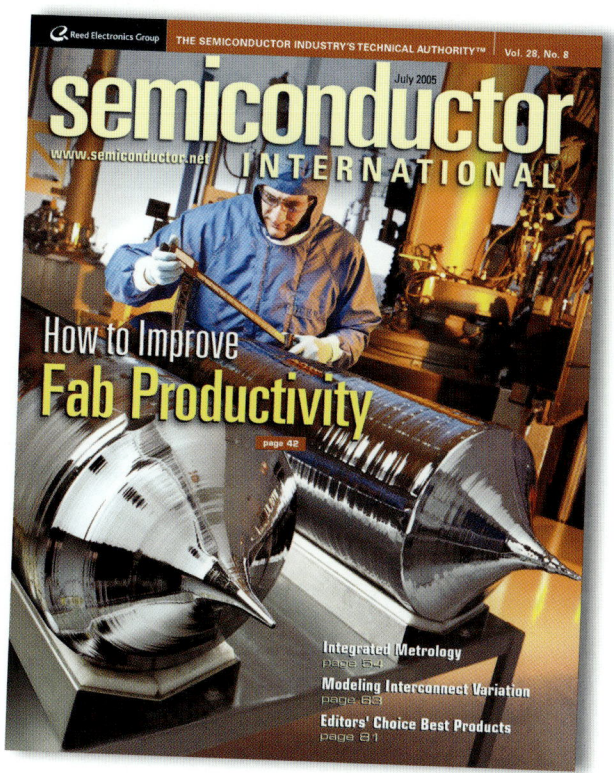

MEMC was featured in the July 2005 issue of *Semiconductor International*.

Timeline

2002 — Worldwide semiconductor materials sales reach $21 billion, exceeding the size of the equipment market for the first time.

2003 — Satellite radio is invented.

2004 — 300mm capacity exceeds 100,000 wafers per month.

MEMC enters into an agreement with Silicon Genesis Corporation to manufacture SOI wafers using their layer transfer technology.

2005 — 300mm wafers go into production in Taiwan.

2006 — MEMC announces $5-6 billion agreement to supply Suntech Power Holdings Company with solar-grade silicon wafers.

Despite the gloomy conditions, MEMC continued to position itself for the time when the market would turn around. In March 1997, the first phase of the 300mm research and development line at St. Peters was completed, and 300mm wafers were being shipped to customers worldwide. Additional investments in the Utsunomiya, Japan, facility, totaling some $250 million, were announced in October of 1997.

TRANSFORMATION

Ludger H. Viefhues was named president and CEO succeeding Roger McDaniel in 1996. He immediately launched a company-wide program called Transformation to shore up MEMC's competitive advantage. Calling on employees throughout the company, the goal of the Transformation was to examine all policies, procedures, and practices on a functional, rather than regional, basis. Teams were organized, locally and globally, and were asked to "stick their oars in the water and row in the same direction."

Customer Focus Teams investigated every aspect of customer service. No detail was too small for examination by Accelerated Cost Reduction Teams (ACRT). Reviews began on such items as the silicon waste that resulted from cutting ingots, improving packaging to minimize the "outgassing" effect on wafers, and so on. Blitz Teams were formed as cross-functional groups whose job it was to focus on one function and, without any capital expenditure, implement cost-cutting measures within five days. Global Best Practices conferences were launched as a continuation of the one initiated by the Cheonan, South Korea, plant back in 1996.

Though these efforts ultimately proved to be productive, the challenges of over-

Bob Sandfort and Roger McDaniel are interviewed by the press at the New York Stock Exchange when MEMC went public on July 13, 1995.

Above: MEMC began trading as a public company under the symbol "WFR."

Right: Unknown, Ludger Viefhues, Roger McDaniel, and Bob Sandfort at the NYSE, 1995.

capacity and resulting price reductions remained. Plans for a joint venture in Malaysia, MEMC Kulim, were cancelled in 1998 as were plans for another joint venture in China. The recession also contributed to the shutdown of the Spartanburg, South Carolina, plant.

Revenues decreased again in 1999, but better days were just around the corner. In 2000, MEMC acquired an additional 40 percent interest in MKC (formerly POSCO Hüls Company, Ltd.), increasing its ownership to 80 percent with Samsung retaining its 20 percent portion. As a result, MKC's operating results were consolidated with MEMC's, giving the company a positive net income in the fourth quarter of 2000, the first since the beginning of the industry recession in 1997. Progress seldom happens in a straight line, however, and continuing problems in the semiconductor industry translated to a 29 percent reduction in MEMC's sales in 2001.

CHANGES AT THE TOP

In February 1999, Ludger Viefhues retired as MEMC's president and CEO. Klaus von

Strictly Business 67

Hörde moved from COO to CEO in his place and continued the processes initiated in the Transformation program.

In June of 2000, Hüls' parent company, VEBA AG, merged with VIAG AG to form E.ON AG and announced its intention to focus on its core business of energy and specialty chemicals. A majority shareholder of MEMC at the time, their intention was to divest MEMC, among other noncore businesses. In November of 2001, an investor group led by Texas Pacific Group (TPG) purchased E.ON AG's interest in MEMC.

In April 2002, von Hörde retired, and Nabeel Gareeb took the helm as president and CEO of MEMC. Gareeb, formerly chief operating officer of International Rectifier, a leading supplier of power semiconductors and system solutions, brought expertise in technology and operations to MEMC along with his unique vantage point as a former customer of the silicon wafer industry.

By this time in 2002, silicon wafer manufacturers had consolidated to the point where four major companies—MEMC, Shin-Etsu Handotai (SEH), Sumitomo Mitsubishi, and Wacker—accounted for more than 80 percent of wafer production. As the market slowly began to show signs of recovery, MEMC, under Gareeb's unbridled determination, reignited its focus on the fundamentals: technological supremacy, meeting and exceeding customer demands, and efficient operations management. This back-to-basics

Group visiting Spartanburg plant, left to right: J. H. Mao, B. Wrather, C. B. Lee, K. Schriewer, L. Hand, P. Lupano, D. Hale, B. Lukas, A. Bertoili, D. Hargett, M. S. Lin, D. Rice, K. Sasaki, D. H. Suh, Lee Wee Loke.

Crystal pulling floor at Spartanburg.

Nabeel Gareeb, president and CEO, 2002–2008.

approach paid off. Net sales at MEMC grew 11 percent in 2002, 14 percent in 2003, and 32 percent in 2004. Within that time period, products introduced in the prior three years represented 30 percent of MEMC's sales, and the *Wall Street Journal* recognized MEMC as one of the ten best performing technology stocks in 2002. Rose Associates, a market research firm based in Los Altos, California, called MEMC's revival the "biggest financial turnaround in electronic materials history."

CONTROL THE COMPANY

In July of 2002, TPG converted its preferred stock to common stock, increasing its ownership of MEMC to approximately 90 percent. Before long, that trend was reversed. From 2003 to 2005, TPG reduced its ownership of MEMC through stock offerings. The final offering in February of 2005 reduced its stake in the company to 34 percent, giving MEMC greater internal control. By the end of 2007, TPG reduced its beneficial ownership of MEMC's common stock to zero.

In 2004, MEMC acquired the remaining share of both Taisil Electronic Materials and MEMC Southwest in Sherman, Texas, having already increased its ownership of MKC to 80 percent back in 2000. That same year, MEMC broke the $1 billion revenue mark and moved up into the number three position in market share.

A PERIOD OF REBIRTH

Throughout his tenure at MEMC, Nabeel Gareeb adhered to a self-funding business model designed to build the company's resources and cash flow generation as security against future industry-wide cycles. Acknowledging that the silicon wafer industry has historically been subject to highs and lows, the self-funding business model was intended to give MEMC the resources it needed to ride future waves and provide the cash resources necessary for the development of future technologies.

In 2002, MEMC began a renewal initiative designed to position the company for the significant growth opportunities that MEMC experts predicted. As Phase I of this initiative drew to a close in 2006, the company proudly reported a five-year period of sales growth at a compound annual rate of 20 percent. MEMC was named to *Forbes*' list of Best Big

Klaus von Horde, president and CEO, 1999.

Strictly Business 69

MEMC GOES INTO OUTER SPACE!

In July 2001, NASA launched the Genesis spacecraft on a journey of a million miles toward to the sun to collect pieces of the sun, called solar wind, in an attempt to answer questions about the sun's composition and its origins.

The start of that million-mile journey began in 1995 when California Tech contacted Dr. Graham Fisher, then director of Manufacturing Technology, and requested a sample of high purity silicon for neutron activation analysis. MEMC's Dr. Greg Allen headed up an applications engineering project to develop the requested silicon. The results showed impurity levels below detection, and tests involving thermal simulations in space showed that MEMC's standard acid-etched surface performed best at radiating heat in a near vacuum —a major concern for solar collectors.

In December of 1997, NASA selected the Genesis Project as one of only two Discovery Mission projects to be funded over the next four years. Approximately 1,100 unpolished wafers were shipped from St. Peters to NASA in November 1998 for laser-cutting into hexagons for use in the project.

MEMC produced these wafers for NASA's Genesis Project.

MEMC regularly hosts educational workshops for customers. Workshops have been held in Belgium, Germany, Taiwan, and the United States.

Kuala Lumpur, Malaysia, plant, 1994.

Utsunomiya, Japan, plant, 1994.

Novara, Italy, plant, 1994.

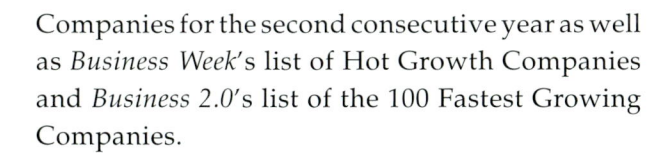

St. Peters plant, 1997.

Companies for the second consecutive year as well as *Business Week*'s list of Hot Growth Companies and *Business 2.0*'s list of the 100 Fastest Growing Companies.

A significant driver of growth in the coming years was projected to come from the sale of 300mm wafers, which had a higher selling price per unit, helping make them a revenue and margin multiplier. In addition, MEMC continued to focus on cost controls and maximizing its return on investment capital. By 2006, MEMC showed improvement in every category of the business model; increased sales, cost reductions, and operating cash flow led to increased return on assets, key to an asset-intensive industry. Though polysilicon remained in tight supply, MEMC's ability to produce 90 percent of its polysilicon requirements internally gave the company an enviable strategic advantage.

Upon his resignation in October of 2008, Nabeel commented on the success of MEMC's business model: "With our polysilicon expansion now past the early part of the learning curve, we have also recently enhanced this strategy by cultivating additional shorter-term wafer customers in the solar space, while continuing to support the growth of our existing long-term customers. I believe the strength of the combination of asset efficiency, market positioning and strong cash generation should provide a springboard for the company to achieve even greater success in the future."

CHAPTER SEVEN
Equations for Success

Lifestyles have changed dramatically in the fifty years since the first shovelful of dirt was moved to make way for the St. Peters plant. The average home has laborsaving devices that would have been the envy of previous generations. Flat screen televisions with the ability to receive over 150 channels have replaced the radios of the 1940s and bulky television sets of the 1970s. Bills are paid and shopping and homework are done on laptop computers with high-speed Internet access. Teenagers, once relegated to their bedrooms to play vinyl LPs on record players, now download their favorite music to their highly portable MP3 players. Today's grandparents communicate with grandchildren via e-mail and receive photos of their latest soccer games on their cell phones. And, today's automobiles may have as many as fifty microprocessors or minicontrollers. All of this was achieved by the invention and continued innovation of the integrated circuit, built on silicon substrate, and continuously enabled by MEMC.

Fifty years into the Information Age, technology is irreversibly embedded in cultures around the world. What has sustained MEMC through the past fifty years will, without question, position it for the future: its technological innovation, its focus on the customer, and the people who make both possible.

> **DID YOU KNOW?**
> By the end of the twentieth century, more than 250 million people were living in the United States and owned approximately 1.6 billion electronic devices.

TECHNOLOGICAL EXCELLENCE

One of the secrets of MEMC's success undoubtedly rests in the tremendous depth of the company's research and development knowledge. The R&D work that MEMC accomplishes each and every year leads to constant

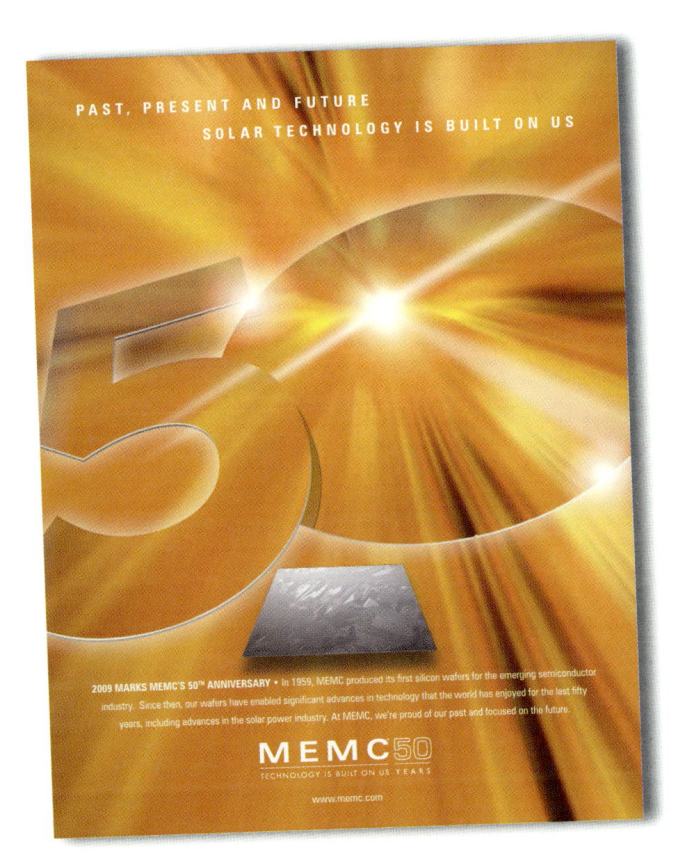

MEMC proudly announces its fiftieth anniversary celebration.

advances in wafer technology and increased customization that allows MEMC broader market penetration. In the last five years alone, MEMC has invested over $187 million in research and development and currently holds more than six hundred worldwide patents on silicon products and processes, as well as an additional three hundred-plus worldwide patent applications currently on file. MEMC's focus on research and development and its long-standing tradition of quality and customer service have culminated in a company that is better positioned than at any point in its history to meet the demands of the highly competitive and fluctuating electronics industry.

In the past five decades, MEMC has played a major role in the explosion of technology in the Information Age. From the early quarter-sized finished wafers (complete with tiny devices) pictured right, to today's multicrystalline solar cells for the solar industry, MEMC is helping to build technology.

MEETING THE CUSTOMER'S NEEDS

Understanding that MEMC's success is dependent upon the success of the customers it serves, MEMC's products are manufactured to their precise and demanding specifications. The combination of varying diameters combined with physical and electrical specifications geared to each customer and application equates to more than 2,200 different wafer specifications and could reach close to 3,000 by the end of the decade. Reflecting this level of application-specific production, John Kauffmann, senior vice president of Worldwide Sales, Customer Service, and Marketing, describes MEMC's sales strategy as one that focuses on "delivering targeted products to strategic partners."

MEMC'S COMPETITIVE EDGE: A TRULY GLOBAL WORKFORCE

One of MEMC's greatest assets is without question its people—a global workforce that brings a variety of problem-solving approaches to the table and a unique ability to interact with customers around the world. They bring with them a drive that fosters innovation, always anticipating the needs of the future semiconductor, and now solar

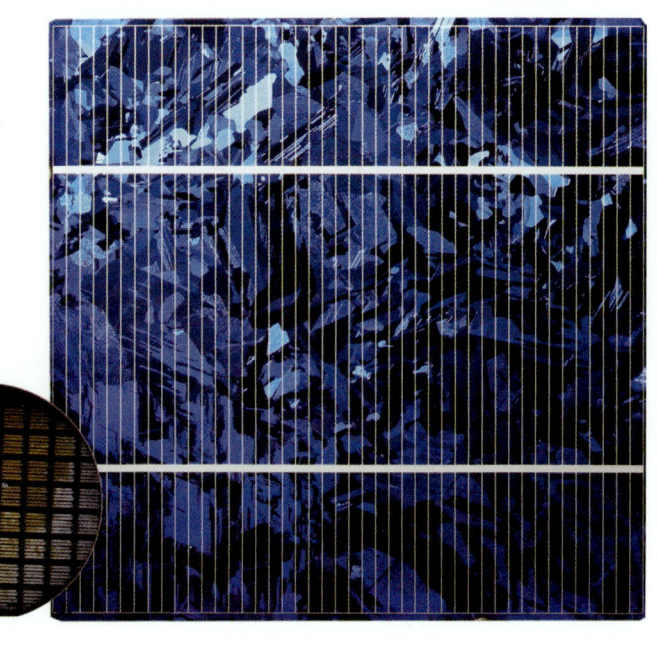

Equations for Success

May 2000
MEMC receives the C. Sheldon Roberts Award of Excellence as a Top Supplier from Fairchild Semiconductor.

February 28, 2001
MEMC marks the shipment of its one-millionth 200mm wafer to Intel.

2003
The AMD Achievement Award is presented to MEMC for World Class Performance.

2004
Intel presents MEMC with an award for its significant contributions to the development of Advanced Si Metrology.

2006
MEMC receives the Fast Track Award from the Missouri Chamber of Commerce, recognizing it as one of the fastest-growing companies among its members.

2007
Frost & Sullivan presents MEMC with their Award for Leadership in Financial Management.

February 21, 2008
MEMC conducts the opening bell ceremony at the New York Stock Exchange.

2008
MEMC is ranked #10 among *Business Week* magazine's list of Top 50 Performers for 2008.

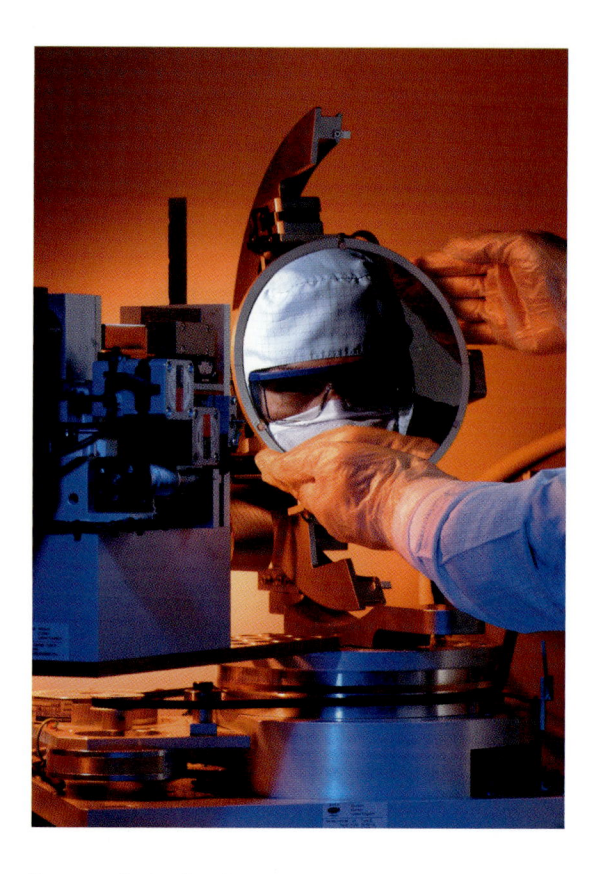

X-ray analysis of wafers.

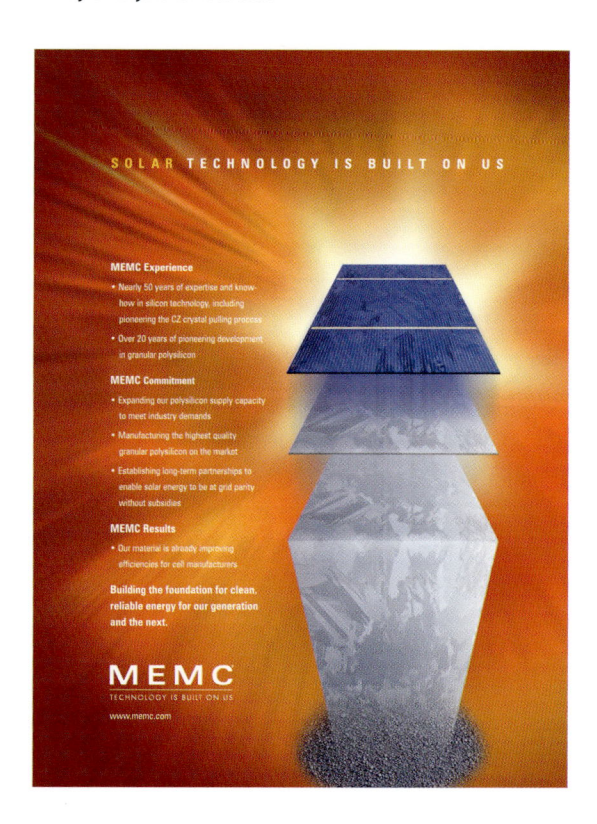

MEMC ad for solar wafers.

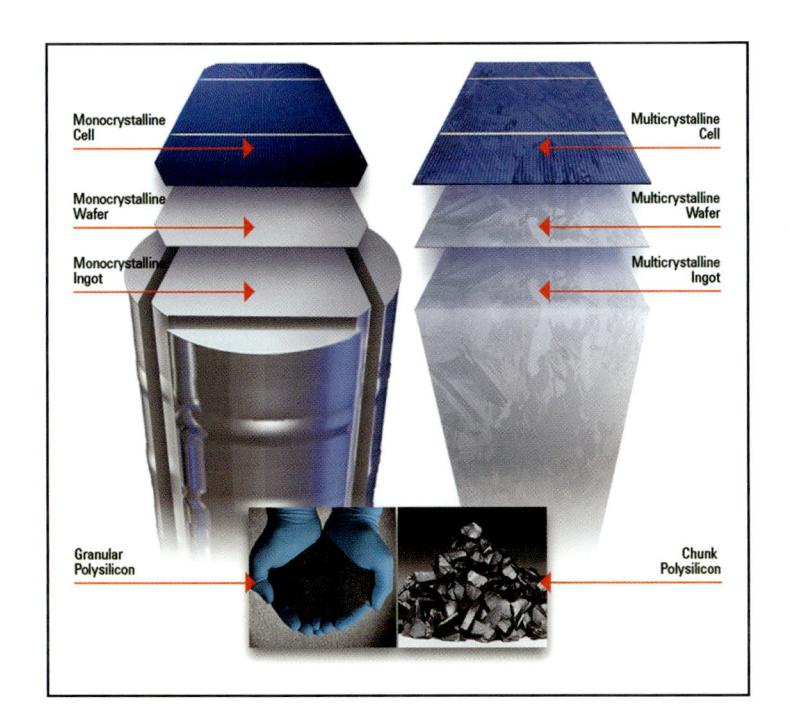

From chunk and granular polysilicon comes finished solar cells.

MERANO, ITALY

While visiting the Merano plant in 2000, Jerry Canfield, who was a St. Peters employee at the time, reported that the major difference between St. Peters and the Merano plant was the food. Meals were served in a cafeteria and included pasta, salad or meat, and vegetables—"no burgers and fries here," he reported. He also noted that instead of the omnipresent American cup of coffee sipped leisurely throughout the day, folks at Merano got their shot of espresso, "slammed it right there in front of the vending machine," and went right back to work.

industries, where advances originate. If the people of MEMC are proud of the job they do (and they are), it is also true that they are proud of their peers and proud to be part of MEMC's very diverse workforce. When interviewed, both John Kauffmann and Dr. Graham Fisher identified the diversity of MEMC's workforce as one of its most important assets. John described his experience in Japan as an introduction "with another whole level of quality" and in Taiwan, "a whole new level of productivity," describing the people of Taiwan as very innovative and hard-working people: "Our strength vs. our competition comes out of that diversity. We have experts essentially in every area of silicon that can speak and communicate in the mother tongue of whatever country we're in and that's not the case with our competition."

Dr. Graham Fisher observes on a regular basis how this diversity impacts MEMC's technological growth: "This company has people from all over the world, bringing with them their varied educational backgrounds, different approaches to problem solving, different ways of looking at the market. The result is a synergy of ideas and knowledge that replaces the narrow assumptions that can occur in a more homogenous workforce."

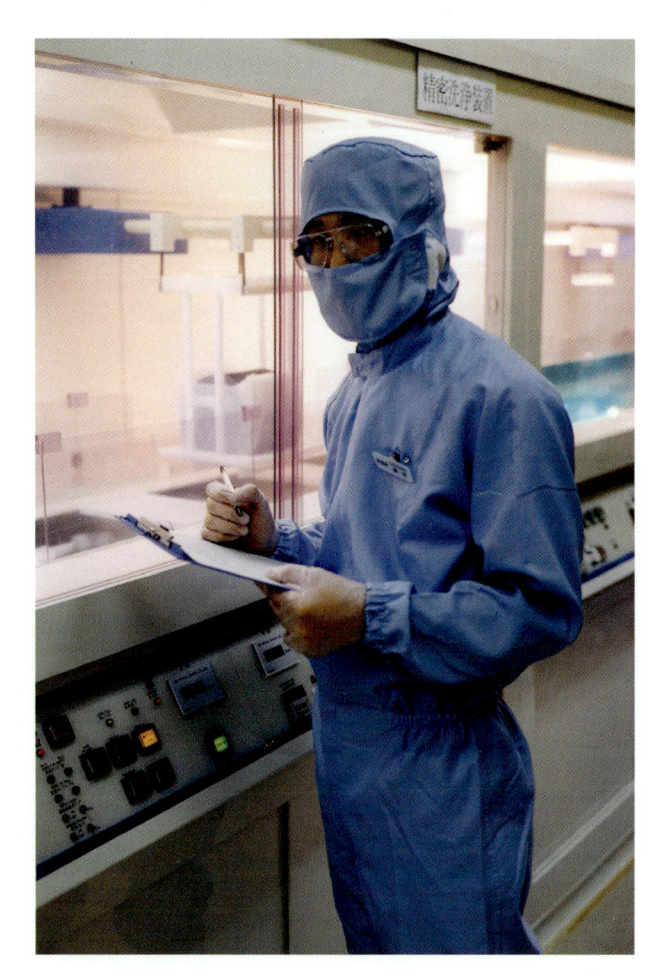

Wafer final cleaning at Utsunomiya, Japan.

Equations for Success

The people of MKC celebrate their culture as well as the work they do at MEMC.

Asian Conference, June 1996.

EXCHANGING IDEAS AND KNOWLEDGE

When MEMC initiated the Transformation program in 1997, meetings were facilitated at every location; brainstorming sessions were designed to get input from everyone to develop the best possible strategy for change. Teams were organized to monitor the change strategies that emerged, and the names given those teams is an indication of how much MEMC values its global workforce: "Global Integration Teams, "Local Area Change Agent Teams," and "Global Break-through Teams" were set up to communicate and monitor the change strategies. Dr. John DeLuca, corporate vice president of technology at the time, said that throughout the process, people "became initiators of formal and informal networks for exchanging ideas and knowledge throughout MEMC."

KUALA LUMPUR, MALAYSIA

1991	Quality Excellence Award by Cherry Semiconductor
1992	Superior Quality Award from Medtronic
1997	Supplier Partnership Award from SSG Thomson
2001	Intersil Supplier of the Year Award

Throughout MEMC's past newsletters you can find numerous examples of how employees at all locations take pride in their ability to contribute to MEMC's success. The varied ways in which this pride is expressed often reflects the culture and customs of the country in which the plant is located, embodying the notion that excellence and achievement are without geographic boundaries.

The Spartanburg plant started "The Perfect Wafer" program in 1996 to increase customer satisfaction and reduce customer advisories by 40 percent every year.

In April of 1997, an audience of more than 450 people gathered as nine teams competed in their performance of a song that had been commissioned by PHC at MEMC's joint venture plant at Cheonan, Korea. The goal was to improve harmony and promote a sense of pride and belonging at the plant. Also at PHC, teams of employees participated in tree planting events as a way of reinforcing the idea that the growth of the company requires a team effort.

In January 1996, twenty-one engineers from six MEMC plants met at a "Best Practices" conference at the Utsunomiya site. The idea came from S. C. Chang while on assignment from Korea. Chang was looking for ways to compare the performance of process on the 200mm wafers between St. Peters

Silicon wafer manufacturing diagram.

and the Cheonan plant. As a result of his visit to St. Peters, five Cheonan plant employees spent a month studying the slicing operations at St. Peters, resulting in a dramatic improvement in yield. The success of that study led to a "best practices" conference in which all 200mm-producing plants participated. These conferences provide an opportunity for a synergistic exchange of ideas, the net result of which is that the best processes from each location are implemented throughout the company.

The worldwide implementation of a crystal pulling machine control technique used in Merano, Italy, led to a 1.5 to 2

TAISIL ELECTRONIC MATERIALS

May 1997	Excellent Company in Labor Working Environment
November 1998	Top 10 Excellent Plants in Industrial Safety and Health
November 1998	Top 10 Excellent Company in Training and Development
July 2000	Outstanding Silicon Supplier Award from UMC VPP certification
April 2002	"Ship to Stock" Certificate from TSMC
September 2002	Best Silicon Supplier Award from TSMC

Equations for Success   77

2009 website www.memc.com

percent increase in wafer yield in 1998. The same year, a process that recovers a majority of the slurry used in cutting crystal ingots, developed in Novara, Italy, was adopted worldwide.

MEMC FELLOWS

Wherever extraordinary effort and talent exists, it deserves to be recognized. To that end, MEMC's Fellow Program recognizes and rewards technical leaders within the company for their contributions to the business success of MEMC. Nominees are limited to those who follow a technical specialization career path instead of a management career path and who have demonstrated outstanding,

sustained performance. Every year, a banquet is held to recognize the MEMC Fellows and celebrate their achievements.

At the start of 2009, there are thirty-six members: seven Senior Fellows, six Fellows, and twenty-three Associate Fellows. Twenty-two are from St. Peters, three from Japan, four from Korea, one from MEMC Southwest, and three each from Novara and Merano, Italy.

EXPATS

Perhaps the greatest commitment to diversity is illustrated in MEMC's "expats." In 1996, MEMC had

MEMC celebrates its fiftieth anniversary by developing a new newsletter for all employees worldwide called *WFR News*.

over thirty expatriates, people who work outside their home countries, but whose compensation and career are tied to their employer back home. While the experience of working in other countries certainly provides one with an opportunity to develop the cross-cultural skills that are necessary for dealing with the complexities of international business, it can also be a trying experience not only for the employee, but for their families as well. Working as an expatriate requires the patience to learn to adapt to the culture one has joined instead of asking the culture to adapt to him. The value, both to the company and the individual, lies in embracing and understanding a culture outside of one's own native culture.

THE FUTURE AT MEMC

When Ahmad Chatila was named president and chief executive officer in March of 2009, he was placed at the helm of a company that is a leading supplier to the almost $1 trillion electronics market

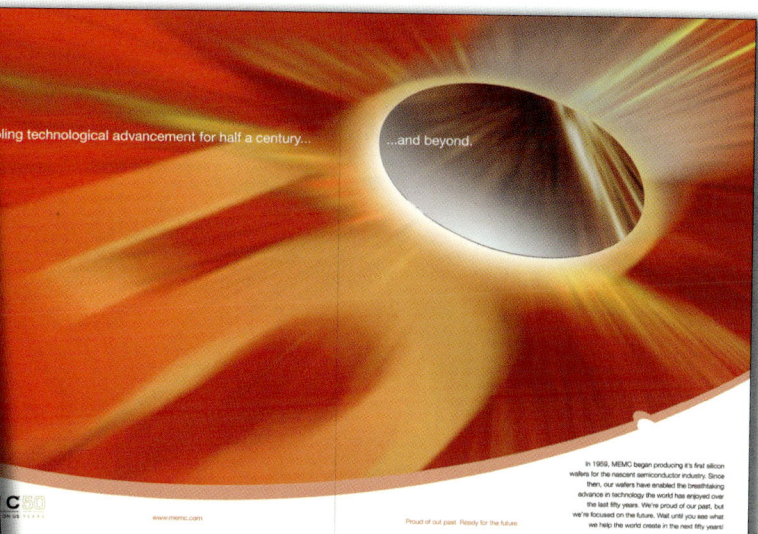

From 1959 to 2009, MEMC has stood for technological excellence.

and the $18 billion solar industry. A global company in every sense of the word, MEMC has locations in the United States, Japan, Taiwan, Korea, Malaysia, and Italy. Employing approximately five thousand people around the world, MEMC's workforce is arguably one of the most highly educated workforces in industry.

No other technological advancement has impacted human history more than that which has made the

MEMC KOREA COMPANY, CHEONAN

1993 & 1998	Best Safety Management
2002	Awarded "Excellent Rewarding Work Places together with Labor-Management" from Korea Employers Federation
2002	Awarded "Man & Woman Employment Equality Excellent Enterprise" by the Ministry of Labor
2006	Best Human Resources Developer Certification from the Ministry of Education and Human Resources and Labor

Equations for Success

Ahmad Chatila, MEMC President and CEO.

Information Age in which we now live possible. The technology made possible through the production of silicon wafers has revolutionized forever how we create, disseminate, retrieve, and store information. The people of MEMC, and the products they develop and manufacture, have made that possible. They have earned the right to proclaim, "Technology Is Built On Us."

IPOH PLANT

The newest and latest isn't always about technology. In 2009, MEMC plans to open a new facility in Ipoh, Malaysia. The new facility will provide a low-cost manufacturing site, which will improve MEMC's ability to penetrate and increase share in the 200mm market. Ipoh will have the capability to perform wafer slicing through epi. It is designed to handle products for all market segments, more specifically to hold a world-class facility to support the growing 200mm discrete and CIS markets.

September 2007	Began property acquisition in Ipoh
June 2008	Commencement of construction to extend, renovate, and convert property into clean room manufacturing area with required facilities
Mid-to-late 2009	Expected to start product qualification of 200mm wafering

Sources:
St. Louis Post Dispatch, Business section, May 19, 1995
St. Louis Post Dispatch, May 18, 2007

MEMC's newest facility in Ipoh, Malaysia, is scheduled to open in 2009.

MEMC World Headquarters, St. Peters, Missouri.

PerfectSOI™ **Perfect Silicon**™ advanta